Thames Television's KITCHEN GARDEN

KEITH FORDYCE
AND CLAIRE RAYNER

Illustrated by
Norman Barber

A
TVTIMES
BOOK

Independent Television Books Ltd, London

CONTENTS

Growing Your Vegetables

Vegetable Recipes

Thames Television's Kitchen Garden

This book is based on the Thames Television series, produced by George Sawford, which was first transmitted in February 1976. Keith Fordyce presented the gardening part of each programme and Claire Rayner, the cooking section.

For help with the book, Thames Television is grateful to Jim Smith of Suttons Seeds.

GROWING YOUR VEGETABLES

KEITH FORDYCE

GLOSSARY

Bed out
to plant a seedling into its final growing place

Brassicas
the word used to describe the cabbage family and some roots such as swedes and turnips

Calomel
a chemical usually used as a 4 per cent compound in a dust to stop club root in brassicas, carrot root fly etc

Compost
rotting vegetation used to fertilise the soil, or a special soil mixture for raising seeds and seedlings

Dibber
a pointed piece of wood used to make a hole for tubers or seedlings

Drill
a furrow made in the ground, which can be either v-shaped (page 14) or flat-bottomed (page 16) in which to sow seeds

Fertiliser
plant food put into the soil to replace that used up by previous crops

Haulm
the foliage of potatoes, runner beans and some other plants

Humus
decayed vegetable matter used to enrich the soil

Inorganic fertiliser
artificial fertiliser as opposed to organic fertiliser, quicker acting than organic fertiliser

Manure
decomposed animal or vegetable matter used as a soil nutrient

Nitrogenous
containing nitrogen, one of the most important requirements for plant growth

Organic fertiliser
a fertiliser consisting of decomposed animal or vegetable matter, ie horse manure, compost, etc.

Pinching out
removing unwanted shoots from a plant

Seedling

a young plant consisting only of the main stem and no side shoots

Singling

to reduce a group of seedlings to one

Sow

to put seeds into a suitable medium to start growth

Spit

the depth of a spade, usually about 30cm (12 inches)

Thinning

removing some seedlings to enable those left in the soil to have room to grow

Tilth

soil broken down very finely with a rake or hoe for sowing seeds

Transplant

to move a growing plant from one place to another, ie from pot to garden

Truss

the stem of the plant that bears clusters of flowers or fruit

Tubers

the name applied to the swollen roots of plants like potatoes or dahlias, these roots are a food store and produce shoots for the following year's growth

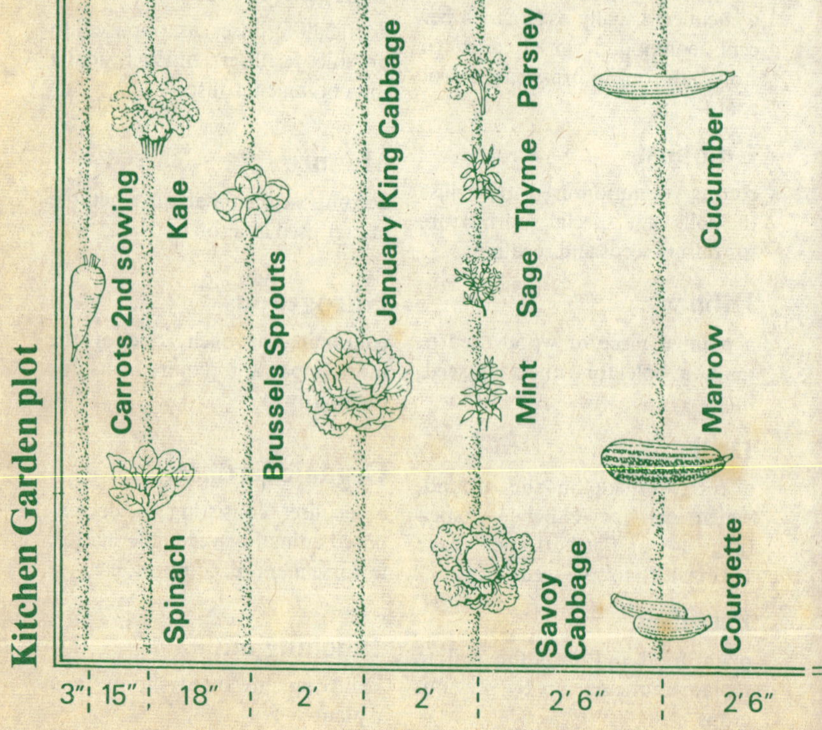

1 Groundwork

In the spring a gardening man's fancy lightly turns to thoughts of double spit digging, compost heaps, manure, and what he will be harvesting in June. When you get the urge to grow something, early spring is the time to start. Even if your sole contribution to the world of horticulture has been a saucer of mustard and cress at the Infants School, don't be disheartened, vegetable growing is not as difficult as you may think.

Choosing the site

The kitchen garden plot is 7m × 3m (24ft × 10ft) that is, about a quarter the size of the garden of the average semi-detached house, and on this plot you can grow over thirty different crops—sufficient to supply most of the requirements of two people for a year. If you are growing for a larger family then increase the size of the plot and the number of vegetables accordingly. The site you choose should get as much sunlight as

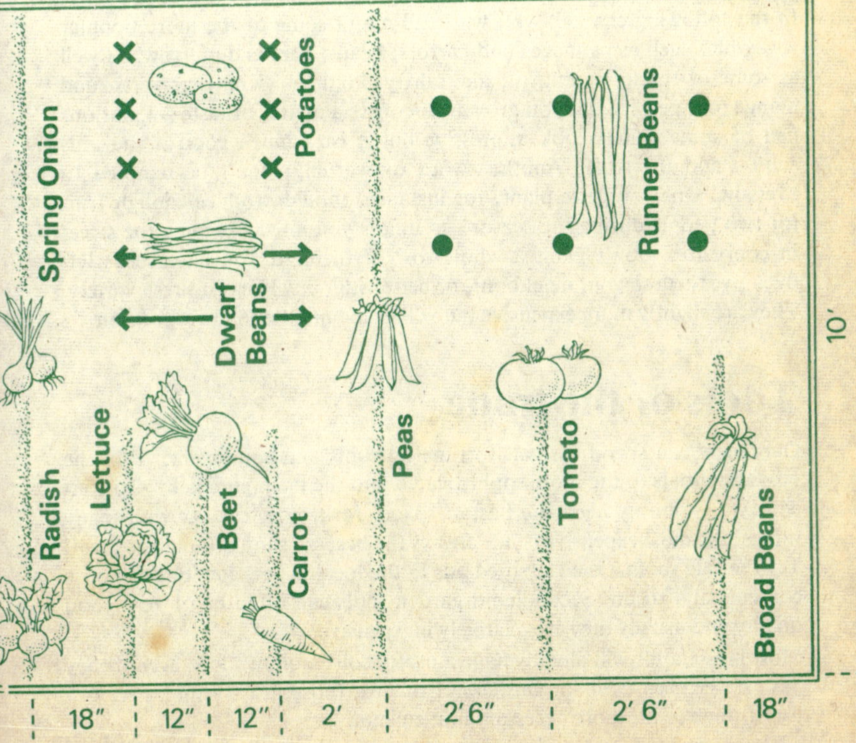

possible and should be sheltered from the prevailing winds. Do not make the mistake of putting your vegetable plants under the old oak tree, because nothing will thrive there. The young plants will simply race upwards in an attempt to reach the sunlight, and finally run to seed.

If your back garden is unsuitable, then why not consider the front? The neighbours may titter at your cabbage rows where the cabbage rose should be but it will be your turn to laugh in the early summer when the neighbours are paying the earth for their vegetables and yours are coming fresh and free. If neither front nor back garden is suitable, and you still want to grow something, don't despair; you can always apply for a council allotment instead. Be warned though, there is a waiting list for allotments, particularly in the larger urban areas, so the earlier you apply the better. An allotment will cost you approximately 16p per week per rod. (A rod is just under thirty square metres—slightly larger than the kitchen garden plot.) Write to the parks and gardens department of your local council for an application form.

What to grow

In the following chapters you will read about some of the more popular vegetables such as cabbages and carrots, beans, beet and marrow, as well as some of the lesser known, like salsify, kohl rabi, chevriers verts, and mange tout peas. What you grow is of course a matter of taste—a glorious bed of spinach may look impressive but it isn't much good if the kids can't stand the stuff! Another word of warning, don't be tempted to oversow. One courgette plant, for instance, should produce enough fruit for two people during the season, so unless you intend keeping the street in courgettes, don't plant a whole row of them. If you have seeds left over give them to your neighbour, or better still, buy Harvest Fresh variety. They are slightly more expensive but will keep until the following season.

Tools of the trade

Gardeners can spend a small fortune on tools and equipment, from the basic essentials to the more sophisticated and the plain gimmicky. To start with you will only need the basics: spade, fork, hoe, etc. As with many things, the most expensive is not always the best, so shop around for them. Try the sale rooms, the classified ads in the local paper, auctions of house contents and so on. Secondhand garden tools can often be of very good quality and usually they are relatively inexpensive.

Most households will already have a useful collection of tools, even if they have never been used for gardening! But for those just starting, here is a list of some of the more necessary implements.

Strong spade	
Strong fork	
Shovel	
Trowel	
Rake	
Draw hoe	used to make 'drills' (small, shallow trenches for planting beans, peas and smaller seeds)
Dutch hoe	essential for keeping the plot free of weeds, and it takes the backache out of weeding
Small hand or onion hoe	used for detailed work such as hoeing in between small young seedlings.

Three items that are useful and that can easily be made at home are a dibber, a measuring rod, and a wooden mallet.

The dibber is a round length of timber with a point at one end. A broken spade handle is ideal, or even a piece of broomstick. It is used for making small holes for bulbs, tubers and for seedlings when transplanting.

You will find a measuring rod invaluable for measuring distances between rows and between individual plants. To make a measuring rod take a piece of 50mm × 25mm (2in × 1in) timber, smooth it off and paint it white. It can be any length you choose, but best to keep it to manageable proportions—roughly two metres (6ft) or so. On one side of your measuring rod mark off 15cm (6in) lengths with heavier markings on every other 15cm marking (the 30cm (12in) markings). Number the metres (or feet) to however long your rod is. As an optional extra every third 15cm (6in) can be marked 45cm (18in)—this then gives you 15cm (6in), 30cm (12in) and 45cm (18in) markings at a glance. Mark out 38cm (15in) spacing on the reverse side of your rod and there you have all the measurements you will need for your plot in an instant.

A mallet is another useful implement. You will need a piece of hardwood about 38cm (15in) long, and 75 or 100mm (3-4in) square. Make a saw mark right round the wood, about 125mm (5in) from the end. Shave down this short end to a handsize grip, and there you have a made-to-measure mallet. Very useful for driving in canes, stakes and line stumps, etc.

There are many more tools that you may find useful, but for the moment those already mentioned will be enough.

Plant food

The ideal food for plants is manure (except where you are going to plant root crops in the first year). But manure does have disadvantages; farmyard manure is very difficult to come by, and the neighbours may well object to a steaming heap of horse dung over the fence as they sit down to their tea. A gardening friend of mine tells me his vegetables thrive on

A *Strong spade*
B *Strong fork*
C *Shovel*
D *Trowel*
E *Rake*
F *Draw hoe*
G *Dutch hoe*
H *Small hand or onion hoe*
I *Dibber*
J *Mallet*
K *Measuring rod*

elephant dung which he gets from a nearby safari park, but then he has very understanding neighbours . . . they are giraffes! However we don't all live next door to the zoo or the local riding stables, so there has to be an alternative to manure. The most readily available is garden peat used with an inorganic fertiliser. 'Hoof and Horn' and 'Bonemeal' are both good and easily obtainable fertilisers. But home-made compost is in many ways the best, and is certainly the cheapest and most satisfying since it is made from refuse and waste.

Compost

You will need a container for your compost to prevent it spilling all over the garden. Mark out a square, roughly 150cm (5ft). Place posts at each corner, and stretch some wire netting between the posts. Old fencing will do just as well, instead of netting, or you could even use an old packing case, open at the bottom, although this would obviously be smaller than the measurements given above. It is simply a matter of making four walls to contain your compost. Remember not to put the compost heap near to your house or near to the neighbours, because it will tend to smell and won't look very pretty either.

Once you have a four-walled frame you can begin to make your compost. Start the heap with a layer of straw or grass mowings, provided of course that the grass has not been previously treated with a weed killer. Then simply pile in garden and kitchen refuse—not just vegetable peelings, but the stuff you scrape off the plates as well, until it is about 30cm (12in) deep. Add another layer of straw or grass mowings and some sulphate of ammonia, or one of the proprietary brands of compost accelerator; then more waste up to about 30cm (1ft) deep, followed by another layer of straw or grass mowings and a sprinkling of sulphate of ammonia. Turn the compost over at four-week intervals and leave the rest to nature. In a few months you will have a heap of delicious humus—the plants will find it delicious even if you don't.

The compost may take a little longer to mature in the winter months because it likes warmth, so from October through to May keep your compost covered by old sacks.

Vegetable growing is a rewarding and satisfying pastime. Some hard work is involved, but surprisingly little. In the next chapters you will find out how to cut hard work to a minimum, and to produce some succulent vegetables that will have cost you next to nothing.

2 Preparing the soil

So there you are armed with garden tools, seed packets, stout hearts and strong backs. Now comes the hard work. The digging. But before you down tools in favour of growing geraniums in a window box, let me assure you that this is the only time you will encounter really hard work in the whole of the gardening year. The secret of good, succulent vegetables is in the preparation of the ground, and that means digging.

Soil comes in many varieties, but for simplicity it can be split into three different categories—light, heavy, and medium. There is no magic formula for finding out which yours is. If you can't tell by looking at it, and quite often you can't, without a very experienced eye, then ask a neighbour, a local seed merchant or nurseryman, or phone your parks and gardens department at the local council.

There are two methods of digging, single spit digging, and double spit digging. If the soil is light, single spit digging may well be enough. If the soil is heavy or medium, then double spit digging is almost essential. A 'spit', incidentally, is the gardening term for a spade's depth, about 30cm (1ft). A double spit is therefore twice that, i.e. two spade's depth, or 60cm (2ft).

Single spit digging

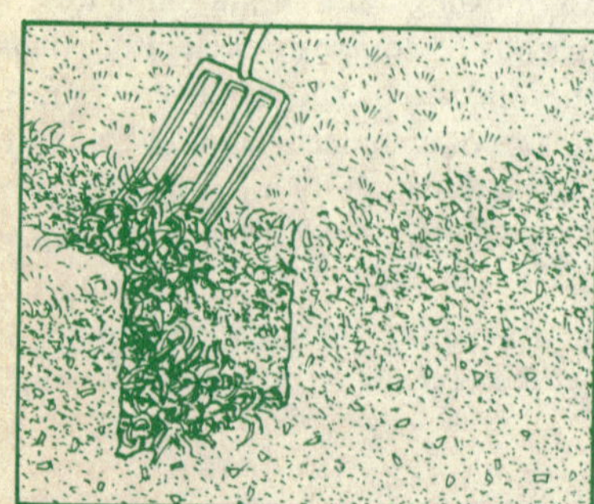

Fork in plant food

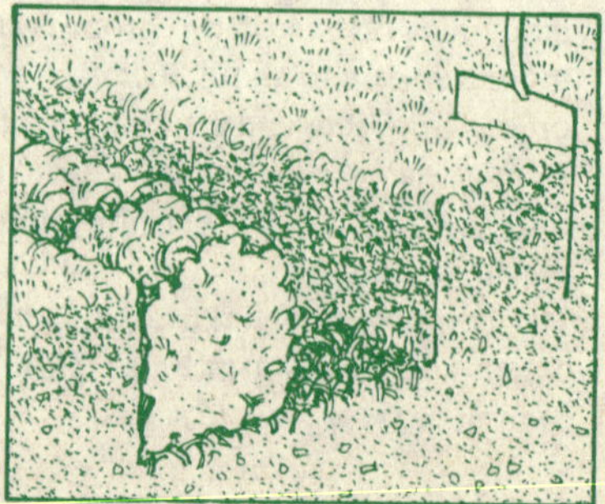

Then transfer soil from second trench to first, repeat

Make a trench across your plot, the whole width of the plot, one spit deep and 46cm (18in) front to back. Take the earth from this plot and place it along the edge of the bottom end of the plot (see diagram). Fork into the bottom of the trench a 5cm (2in) layer of food for the plants. This food should be manure or Hoof and Horn (alternatively Bonemeal), together with garden peat or, better still, your own garden compost (see Chapter 1).

Double spit digging

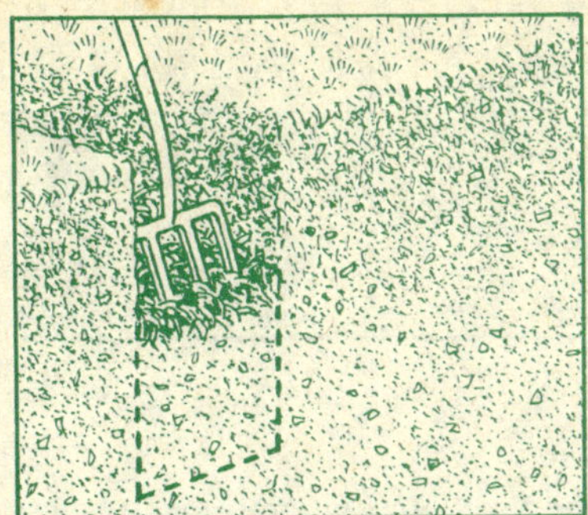

Transfer soil from second trench to first, repeat

Fork in plant food then dig second spit depth

Place your manure, peat or compost in the bottom of the single spit trench, but this time turn over the earth and feeding agent with a fork to a further depth of 30cm (12in), ie another spit. You will need to stand in the trench to do this. All soil benefits from double spit digging, but the lighter soil will need the treatment only once every three years.

Once you have forked in your feeding agent, proceed as follows whether for double or single spit digging. Dig a second trench parallel with the first, filling the first trench with earth from the second. Place your fertiliser in the second trench as before and fill that in with earth from trench three. Proceed in this manner until you reach the end of your plot, where you will find the earth from the first trench waiting to fill the last one.

The above method—taking earth from one end of the plot to the other—requires the use of a wheelbarrow. Do not buy one just for this purpose. Borrow one or else dig your plot in a slightly different way. First divide the whole plot into two equal parts, lengthwise, so that you have two halves, 7m × 1.5m (24ft × 5ft). Dig the first trench only to the half way mark, ie to 1.5m (5ft) across, and put the earth from this trench on the front edge of the other side of the plot (see top p.12). Continue digging this side of the plot with either double or single spit, whichever is appropriate, placing your manure (or forking it in if using double spit). When you get to the end of the first half of the plot, start digging the second half from the bottom upwards, putting the earth from the first trench on this side into the last trench of the first side. Dig the plot back to the top end where, waiting for you, you will find the earth from the first trench of the first side.

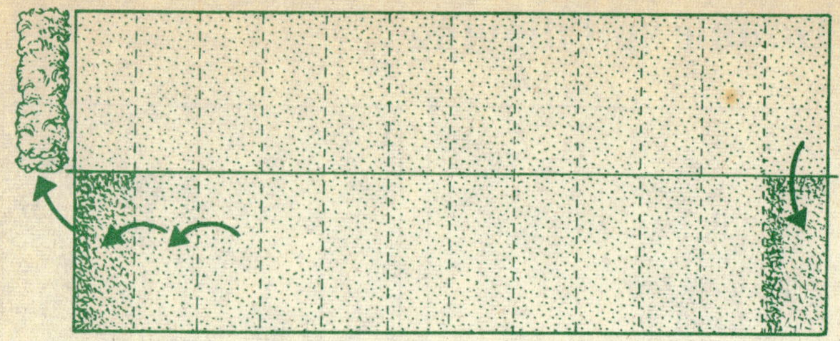

Preparing a tilth

It is unlikely that you found digging and soil preparation to be the most enjoyable pastime, but if you survived your double spit dig, you can rest content that the hard labour is over until next year.

Before putting in the seeds there are two important things to do. One is to rake over your soil until it becomes a very fine tilth—the fine top soil in which you will sow your seeds. If you are able to dig the ground over in the autumn, then the winter frost will save you a lot of work by breaking down the soil for you. If on the other hand the soil is dug in February, then you will need to put in a lot more work. You will need to hack with the draw hoe to break down the lumps of earth, and you may need to tread with your feet to create the required tilth. Always try to dig over your plot in the autumn, as it will save so much work the following season. Remember never do a job yourself if nature will do it for you.

Marking the plot

Now for the second important job. Once you have a good tilth, and before you put in your seeds, mark out your sowing layout. In an area as small as the *Kitchen Garden* plot, 7m × 3m (24ft × 10ft), it is vital to plan ahead in order to avoid wasting space. Using your measuring rod, mark out each row paying particular attention to the distance between the rows (see pages 4-5). If the crops are too close they will crowd each other out and not become fully developed. If they are sown too far apart, you will be wasting precious space, and with a small plot you cannot afford to waste space.

The vegetables placed in our kitchen garden plot are the more popular varieties, but of course you may choose instead any others you prefer—sowing details for the popular vegetables and the rarer varieties are in the next chapters.

3 Sowing early seeds

Once the hard work of digging and soil preparation is over, and the plot is marked out in suitable rows, then you can start thinking about sowing the seeds.

Arm yourself with a good seed catalogue: many of them are free, written by experts and provide an excellent 'read', as well as telling all you need to know of the characteristics of the various seeds you want to plant. Choose vegetables which will prosper in your type of soil.

Usually there will be far more seeds in the packet than is required for the initial sowing. Heavy sowing will result in weak seedlings and great difficulty in thinning out (see Chapter 4). Even if you are lucky, with plenty of sowing space, little and often is the rule, to keep a good succession of young, fresh crops, and to avoid having far more than you can use at any one period.

Having read your catalogue, send for the seeds *early*. Don't delay sending off for them until the day before you need to sow them. And when the seeds arrive, read the back of the seed packet carefully—you will find sowing instructions there that have been written by experts. Always allow for slight variations in the sowing depth. For instance, light soils will probably need slightly deeper sowings.

I give in this and the next chapter details of how to grow the various vegetables. The order is roughly chronological, starting with the earliest sowings in this chapter and ending with the latest in the next.

Broad beans

The first of your seeds to go into the plot are broad beans. This is one of the earliest of the new season's crops, and in the early summer when they are young and at their best they make a welcome change from the winter greens of the earlier months.

Planting out

Broad bean seeds are rather large and therefore need to be sown quite deeply. For a single row, you will need to make a v-shaped drill about 10cm (4in) deep. Mark out the drill, using a line for guidance. This will keep the row neat and straight and will also help to avoid wasting useful space. Take the draw hoe and 'draw' it through the soil a few times, parallel to the line. In drawing the drill try not to put too much of the displaced earth at one end or the other. It should be evenly distributed. The beans should be planted when you have achieved a depth of about 10cm (4in) with the draw hoe. Place the beans about 5cm (2in) apart along the drill. This may seem too close for comfort, but you will be thinning them out later to about 10cm (4in) apart. This is not as wasteful as it may seem. You will still have more seeds left in the packet than you can possibly use, and by planting twice as many as you actually need, you will give yourself a useful insurance policy against the hazards of nature and predators such as birds and mice and slugs. They get hungry too, and will enjoy a good meal from your seeds. Incidentally it matters not a jot which way the beans go into the drill. Whether they are put in flat or sideways or on end, nature has her own way of coping with human cack-handedness, and no matter which way you have put them in, the shoots will come up and the roots will go down!

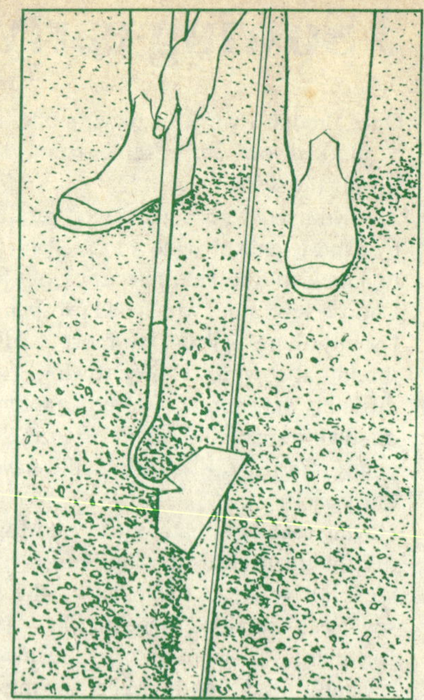

Making a v-shaped drill

Of course, the plot should have been carefully marked out beforehand to avoid wasting space, but if you haven't already marked out the broad bean row, then mark it *before* you fill in the drill, otherwise when you return to your plot, maybe in two or three days' time, you will have to guess where you planted your beans and your guesswork could be out by 30cm (1ft) either way. You could finish up either wasting space, or worse still, by digging up the beans you have just planted! Once you have marked the drill it is safe to fill it. Push the earth into the drill with the hoe, covering the beans, and finally give the ground a light raking over to prevent any hard pans forming.

Pot planting

February, the month in which broad beans should be sown, can be unpredictable, with glorious spring sunshine one year, and below zero frosts the next. As a general rule it is usually safe to plant beans out of doors in February in the South of England, unless the weather happens to be singularly inclement, but what happens if you live in the North? What if the weather is wet and cold? Do not despair! Broad beans, and many other crops can be started in pots or boxes inside the garden shed, or some other

Pot planting

frost-free place. Choose a 10cm (4in) pot, and place in it either a proprietary brand of potting compost, or a mixture of your own soil and peat, firm the soil down and push the bean well in. Fill the pot, leaving room for watering. Beans in pots should be well watered initially and kept in a dark, frost-free shed.

Beans can also be grown in boxes. Place some stones or broken pieces of pot in the bottom of the box to assist in the drainage, and plant the beans about 5cm (2in) apart.

Whether they have been started in pots or boxes, keep the beans in the dark shed until their shoots start breaking through the surface of the soil, then transfer them to a much lighter place. Later when they are about 75-100mm (3-4in) high they will be ready to plant out, and hopefully by then the weather will have improved and will no longer be a threat to young seedlings.

Potatoes

Potatoes are the next to go into the plot. In the South of England they can be planted in March; in the North, perhaps slightly later. Potatoes take up a great deal of plot space, and they are readily available in the shops twelve months of the year, albeit not always that cheap. Therefore plant only a few tubers, simply to provide an early crop of new potatoes at a time when they are at their most expensive in the High Street. Economic reasons apart, what could be better than a dish of buttered new potatoes with a sprinkling of parsley in less than an hour from digging them out of the ground?

You will need only two short rows, 60cm (2ft) apart in which to plant the potatoes (plot plan p.4). Plant the tubers 30cm (1ft) apart in the rows, three to each row giving a total of six plants. It isn't always easy to buy tubers in such a small quantity, so look out for a neighbour who may also be planting potatoes. He may be able to spare you a few of his tubers, or you could perhaps share a bag with some friends. Begged or borrowed, no matter how you come by them, it is a definite advantage to store them in a box in a light cool place—a windowsill is ideal—until it is time to plant them. The light will encourage some green shoots to appear and the plants will then mature more quickly.

To plant the tubers, make a hole with the trowel about 15cm (6in) deep and 75mm (3in) across. Firm the bottom of the hole and place the tuber in the hole with the shoots uppermost. Fill in the hole and firm down the soil. There is no need at this stage to pile the earth up over the potatoes in the conventional method (see page 33).

March sowings

There are a number of crops to be sown in March. Generally speaking, carrots, parsley, peas, spring onions, radishes and spinach go into the plot at this time of year together with a small amount of sprout seed to produce plants for transplanting to permanent positions in early or mid-May. It is difficult to be specific about planting times as they will vary according to weather conditions and also to the part of the country in which you are living. Dates given for planting either in this book or on the back of a seedpacket should be regarded merely as a guideline, not as holy writ.

It is reasonable to assume that anything grown in the South of England will be up two to three weeks in advance of anything grown in the North. Similarly, all crops are delayed after a prolonged winter or unseasonable spring. If you are in doubt about when to sow in your particular area then once again try your parks and garden man, the local nursery or seedsman.

Peas

Peas are another large seed, and they are sown in a flat-bottomed drill. On the *Kitchen Garden* plot this was a full 3m (10ft) long row, the full width of the plot. It is again advisable to mark this out with a line to help keep the row straight. The drill needs to be 75mm (3in) deep and 75-100mm

Making a flat-bottomed drill

(3in-4in) wide, and as this is roughly the width of the draw hoe, it is fairly easy to make. Pull the hoe along the line to keep the line taut and stretched. Scatter the peas into the drill so that they end up 25-40mm (1-1½in) apart either way. Cover in the drill and with a very slight ridge drawn up over the drill you can prevent excessive rain from building up directly over the peas, making them less liable to rot.

Sowing smaller seeds

Smaller seeds do not need to be sown so deeply. Generally speaking, the smaller the seed the shallower the drill, but as regards drill size and other sowing suggestions, always be guided by the instructions on the back of the seed packet and also allow for the nature of your soil. Many of the kitchen garden seeds are of the smaller variety. Brussel sprouts, cabbage, kale, carrots, lettuce, radish, spring onion and parsley are all sown in a similar way.

Parsley

Let's take parsley as an example. Here are sowing details for that seed. Draw a drill 25mm-40mm (1-1½in) deep. Pour some of the seeds from the packet into one palm, and then with the thumb and forefinger of the other hand, distribute the seeds gently and evenly along the drill. Parsley can be sown fairly thickly, as it will get frequent thinning through use.

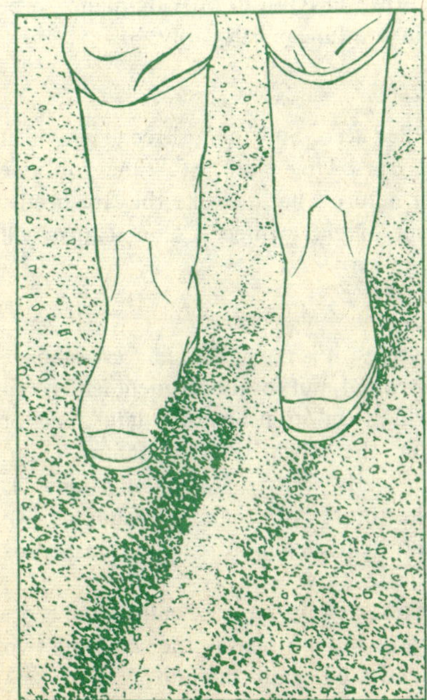

Returning the soil to the drill

The method used to fill the shallow drill is entirely different from that used to fill the deeper type. This time you will need to 'edge' the earth into the drill with your feet, using a method that looks part way between a military two-step, and a dance some readers may remember called the Creep. The idea behind this curious soft-shoe shuffle is to return the soil to the drill without disturbing the tiny seeds you have just put in, and to firm the earth down to exclude any air pockets, without making the ground as hard as cement in the process. Stand with your feet apart on either side of the row and slowly shuffle and twist your toes in and out, edging the soil back into the drill as you do so. Stop every metre (every few feet) and gently tap down on top of the drill with one foot while keeping the weight of the body on the other. Make it a very gentle operation; remember the seeds in that drill are very very small. When the drill is finished and firmed down, give it a finish with a light raking over.

Carrots, radishes, lettuces and spring onions

Other small seeds for March sowing are treated similarly. Carrots, radishes and lettuces should be sown less thickly, since you will need to thin them out later in the growing season. On the other hand, you should not need to thin out your spring onions which are thinned automatically as you pick out the young ones to eat. When planting spring onion it is advisable to dust the drill very lightly with calomel dust, 4 per cent, to take care of the root fly.

April sowings

Dwarf beans and beet go into the plot in April. A small amount of cabbage and kale seed may also be sown in April to provide plants to be set out in their permanent positions in late June or early July.

Beet

Beet is a larger and more easily handled seed. Sow them three at a time at 5cm (2in) intervals along the drill. By sowing three at a time you give yourself an insurance policy against nature's hazards like the flea beetle, and you can afford to lose two out of every three. Calomel dust will take care of the root fly.

Cabbage

Cabbages could rate a book to themselves. It is possible to harvest cabbage of one sort or another the whole year round, but for the moment let us concentrate on Savoys for April sowing, and transplanting late June or early July. Sow in seed beds in the open ground 15mm (½in) deep from the end of April to early May. Transplant them as soon as they are large enough (more of that in Chapter 4). Once again your seed bed will need a light dusting of calomel dust—this time for protection against the cabbage root fly.

Cabbages and other brassicas can be affected by a disease called club root. The best way to avoid this is by not growing brassicas in the same soil two years in succession. Move them around a bit. Don't be put off gardening altogether by a mention of these pests and diseases. Many of them sound far worse than they really are, and most of them are not insurmountable problems. If you have healthy plants, they will fight off most attacks.

Dwarf beans

Dwarf beans should be sown late April—early May. Watch out for late frosts with these, and if you are in any doubt about the weather, leave them for a few more days, because nature has her own way of catching up. Plant the beans in 50mm-75mm (2-3in) deep drills 50mm (2in) apart and then proceed as for broad beans (see pages 13-15).

Runner beans

Shortly after sowing dwarf beans, say mid-May, you will be able to sow the runner beans. It is advisable to put stakes into the ground first (see pp 31-32), before sowing your beans, as this saves disturbing the roots when the beans start to grow. (This applies incidentally to any crop that requires poles or stakes.) Plant four or five beans at the foot of each pole, thinning out to two per pole later when the plants are about 10cm (4in) high.

May sowings

Marrows, courgettes and cucumbers

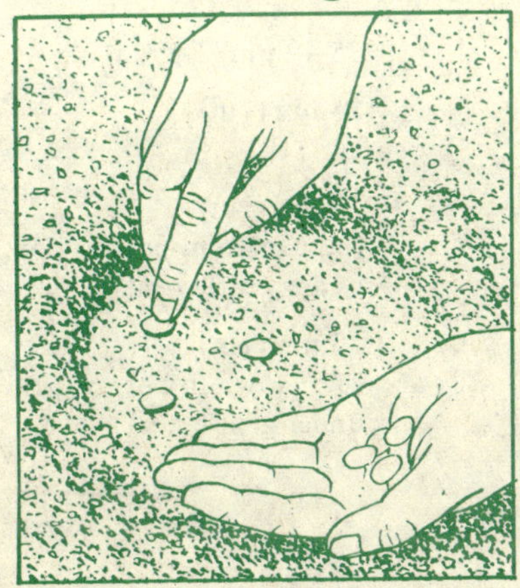

A circular drill

The weather should be getting warmer in May, making it possible to plant the 'fruity' vegetables in the open—the marrows, courgettes and cucumbers. In the kitchen garden plot, aim to have one plant only of each variety. They will provide considerable produce for two, but if you want to keep the neighbourhood supplied with marrows then simply plant a few more seeds. For sowing, it is best to make a circular drill, a shallow depression 13-15cm (5-6in) across and about 4-5cm (1½-2in) deep with a nice flat bottom. The seeds are put in 5cm (2in) apart round the drill, that is, in a circle, and one in the middle for luck. Cover the drill with earth, gently firm down. When the plants are about 75mm (3in) high, choose and retain the strongest one and pull out the remainder.

Thyme, sage and mint

Thyme, sage and mint are best bought as plants. Mint can usually be begged from a neighbour, who invariably will be only too happy to give you some. Left to itself, unthinned mint could take over your plot and in no time at all it would become almost a weed.

SOWING AND PLANTING GUIDE

VEGETABLE	SOWING TIME	DEPTH OF DRILL
Beans, broad	November-April	100mm (4in)
Beans, Chevrier Vert	April-June	75mm (3in)
Beans, dwarf	April-June	75mm (3in)
Beans, runner	May-June	100mm (4in) deep 150mm (6in) diameter
Beet, globe	April-July	40-50mm (1½-2in)
Brussels sprouts	March-April	25mm (1in)
Cabbage, Savoy	April-EARLY May	25mm (1in)
Carrots	March-July	25-40mm (1-1½in)
Cauliflower	May	25mm (1in)
Celery, self-blanching	March-April	25mm (1in) (best sown in pots or boxes under gentle heat)
Celery, trench	March-April	25mm (1in) (best sown in pots or boxes under gentle heat)
Courgettes	May-June	40-50mm (1½-2in) 150mm (6in) diameter
Cucumber	May	40-50mm (1½-2in) 150mm (6in) diameter
Garlic	February-March	25-40mm (1-1½in)
Kale	April-May	25mm (1in)
Kohl rabi	April-July	25mm (1in)
Leeks	March-April	25mm (1in)

SPACE BETWEEN ROWS	TRANSPLANTING if applicable	FINAL SPACE BETWEEN PLANTS
60cm (2ft)	–	10cm (4in)
45cm (1½ft)	–	10cm (4in)
45cm (1½ft)	–	10cm (4in)
120cm (4ft)	–	KEEP 2 strongest plants
30cm (12in)	–	5cm (2in)
60cm (2ft)	May-EARLY June	38cm (15in)
60cm (2ft)	When plants are of planting size	38cm (15in)
30cm (12in)	–	5cm (2in)
60cm (2ft)	When plants are of planting size	38cm (15in)
Plant in a square block	May-June	23cm (9in) each way
120cm (4ft)	June-July	TRENCH 45cm (18in) deep; 45cm (18in) wide; 30cm (1ft) between plants (to take manure or compost). Fill to 30cm (12in) for planting
75cm (2½ft) from adjoining crops	–	Single to best plant
75cm (2½ft) between plants	–	Single to best plant
38cm (15in)	–	10cm (4in)
60cm (2ft)	When plants are of planting size	38cm (15in) between plants
30cm (12in)	–	75mm (3in)
45cm (18in)	When 15cm (6in)	15cm (6in) apart

VEGETABLE	SOWING TIME	DEPTH OF DRILL
Lettuce	March-July	25mm (1in)
Marrow	May-June	40-50mm (1½-2in) 150mm (6in) diameter
Onion, spring	March-May, August-September	25-40mm (1-1½in)
Onion, sets	February-March	25-40mm (1-1½in)
Parsley	March-July	25-40mm (1-1½in)
Parsnips	February-May	25-40mm (1-1½in)
Peas	March-July	75mm (3in) deep 100mm (4in) across
Peas, mange-tout	March-July	75mm (3in) deep 100mm (4in) across
Potatoes–early	February-March	150mm (6in)
Pumpkin	May-June	40-50mm (1½-2in) 150mm (6in) diameter
Radish–salad	March-September	25mm (1in)
Radish–winter	July	25mm (1in)
Salsify	April-May	25-40mm (1-1½in)
Scorzonera	April-May	25-40mm (1-1½in)
Shallots	February-March	50mm (2in)
Spinach–annual	March-September	25mm (1in)
Spinach–perpetual	March-July	50mm (2in)
Thyme	March-April	40mm (1½in) 150mm (6in) diameter
Tomatoes	–	–
Turnips	July	25mm (1in)

SPACE BETWEEN ROWS	TRANSPLANTING if applicable	FINAL SPACE BETWEEN PLANTS
30cm (12in)	–	15cm (6in); 30cm (12in) depending on variety
75cm (2½ft) between plants and from adjoining crops	–	Single to best plant
30cm (12in)	–	Thinned by eating
38cm (15in)	–	75-100mm (3-4in)
30cm (12in)	–	Will thin with use
45cm (18in)	–	15-20cm (6-8in)
60cm (2ft)	–	–
90cm (3ft)	–	–
60cm (2ft)	–	30cm (1ft) (early varieties) 38cm (15in) (late varieties)
135cm (4½ft) clearance all round	–	Single to best plant
25-30cm (10-12in)	–	Thin to single line
30cm (12in)	–	15cm (6in)
38cm (15in)	–	20cm (8in)
38cm (15in)	–	20cm (8in)
30cm (12in)	–	15cm (6in)
30cm (12in)	–	22cm (9in)
45cm (18in)	–	15-20cm (6-8in)
–	–	–
90cm (3ft)	Early June	38cm (15in)
30cm (12in)	–	10cm (4in)

4 Thinning, transplanting and later sowings

As the season progresses into May and June, many of the seedlings will need to be thinned and singled. Some, like cabbages, will need to be transplanted to their final growing place. Around this time you should also purchase tomatoes for planting out.

Fortunately not everything in the garden will need thinning out. Parsley and spring onion will automatically have been thinned out as you picked it for the table. There are three golden rules for thinning and singling. First, choose a cool cloudy day—and that shouldn't be too difficult in England. A recent shower of rain can be a great advantage, because it goes a long way towards helping the seedling left in the ground to recover. Secondly do not try to hurry the job. Take time and do it carefully, remembering that those plants left in the ground will be providing the kitchen with vegetables for the rest of the year. The third rule is to finish the job with gentle hoeing. And one more rule—don't forget to put the thinned-out rejected vegetables on the compost heap!

Radishes and small plants

To thin radishes and indeed all small plants, place the hand upside-down

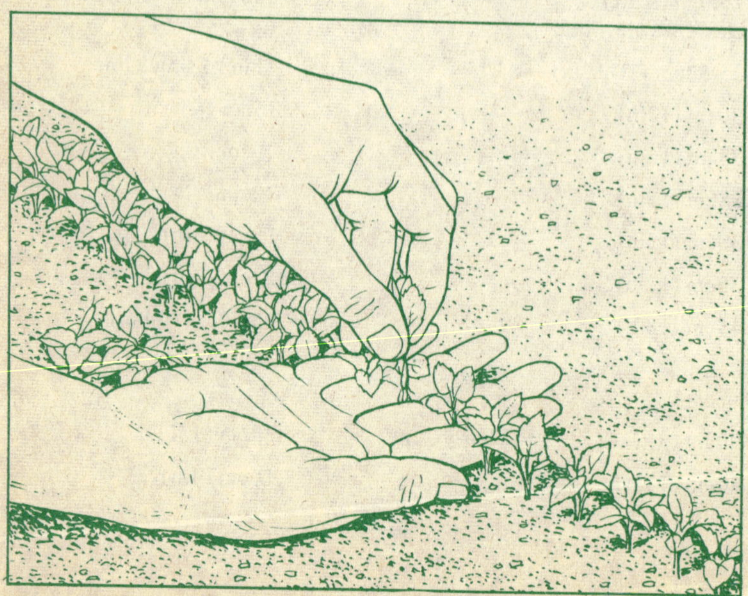

underneath the plant with the back of the hand on the ground to keep the hand steady, then gently tease out the unwanted seedlings with the fingers. The radishes are thinned out to a single row taking care not to disturb those that remain. Some of them may fall over or look rather sad. There is nothing to worry about, they should perk up within twenty-four hours. When you have thinned out sufficiently, give the ground a careful hoeing with a dutch hoe, hoeing the soil on either side of the row. The idea behind this can be compared to tucking a baby back in bed—you've disturbed the blankets and your plants need the warmth and security of the bedclothes, so tuck them up again. It is all a very gentle process, filling the air pockets and helping the seedlings that remain to stand up for themselves again.

Singling lettuce

Lettuces are singled rather than thinned. Eventually they will need to be 15-22cm (6-9in) apart, but for the time being thin them to a 5cm (2in) spacing first as a protection against failure. Then, in about two weeks you will be able to thin them yet again to a final 15-22cm (6-9in) apart in the case of cabbage lettuce. Cos lettuces should be kept 30cm (12in) apart at the second thinning.

Groups of three thinnings

Another type of thinning is the one used for beet, or any other crop that was sown in multiples of three. When the time comes to thin them out you can afford to lose all but the strongest plants. You will need to determine by eye which one of the set of three plants is the strongest and remove the other two, leaving yourself with a singled out row of really strong young plants that will thrive. Once again, hoe the ground after the thinning operation is complete.

Thinning carrots

Carrots are thinned in the same way as radishes. The end result should be a single row of carrots 5cm (2in) apart. Once again like the radishes they may tend to lean over after thinning, but don't worry, give the ground a gentle hoeing after the thinning and in a matter of twenty-four hours the carrots should be fighting fit again. Calomel dust comes into its own again here. Sprinkle some on either side of the row after thinning to prevent an attack of the carrot root fly. Some experts claim that the carrot root fly is attracted by the smell of freshly dug carrots, and that to avoid an attack one should sow the seeds sparsely to start with so that you do not need to thin them, and therefore you diminish the possibility of an attack of carrot root fly.

Thinning beans

The broad beans were planted 5cm (2in) apart. By the time they are about 75mm (3in) high they should be thinned to 10cm (4in) spacing, similarly for the dwarf beans. The runner beans were planted in groups of four or five around the bean pole, and they should finally be thinned to two beans per pole, selecting the best pair from the groups of seedlings at the base of the pole.

Transplanting cabbages, brussels sprouts and kale

The cabbages, brussels sprouts and curly kale will now be looking rather crowded in their seed-bed. Lift them, all very carefully, for transplanting to their final growing place. It is very important to select only the best for transplanting. The seedlings should be about 12-15cm (5-6in) tall. The stems should not be hard and woody, nor, on the other hand, should they be soft and floppy, nor long and thinly drawn out. Choose those which are firm stemmed and close-leaved. Having chosen your plants, you must prepare the ground for them. Place the measuring rod along a line, and dig small holes at 60-75cm (2ft-2½ft) (38cm (15in) spacings between rows if planting more than one row). Fill these holes with water and allow the water to drain away. Place the plant upright in the hole, fill the hole with earth round the plant, and firm the earth in. It is most important that the earth should be well firmed at the root, and you should heel the earth in well all the way round. The acid test is to try to pull out the seedlings afterwards. If they are properly heeled in you should be able to pull the leaves off the plant before you can pull the plant out of the ground.

Leeks, celery and celeriac

Leeks, celery and celeriac are some of the optional extras that are not in the basic kitchen garden plot but none the less are certainly worth consideration. Sow these seeds following the instructions on the seed packets, or in the case of celeriac it may be more convenient to buy the seedlings from the nursery man.

Transplanting leeks

When the time comes to transplant, the leeks should be about 12-15cm (5-6in) high. Before putting them into the soil it is advisable to trim 25-50mm (1-2in) off their tops, as this will help to prevent them from flopping over like wilting lilies and the tips becoming trapped in the

Transplanting leeks

soil. You will need a 10cm (4in) deep drill for the leeks, in fact a mini-trench. Make holes at the bottom of the trench with a dibber, about 15cm (6in) apart—the depth of the hole is determined by the amount of leek that was previously buried (normally speaking that is the white part of the leek). Plant the leeks in the holes and firm them in. This 10cm (4in) drill, or mini-trench, serves a double purpose. In the first place it helps with the watering since a deep trench remains moist for longer. Secondly, as the plant grows, the earth can be teased into the trench to keep the edible part of the leek nicely covered and white, or 'blanched'.

Transplanting celery

Celery has certain similarities in its transplanting to leeks. This time you will need a trench 45cm (18in) deep and 45cm (18in) wide (hard labour again) into which you fork a layer of manure. Place the celery plants in the trench about 30cm (1ft) apart. As with the leeks the soil from the sides of the trench should be gently teased in as the plant gets bigger in order to keep the celery blanched. Experts claim that this type of celery is far superior to the self-blanching type, but for simplicity of cultivation, the self-blanching celery should not be ignored. For this type you do not

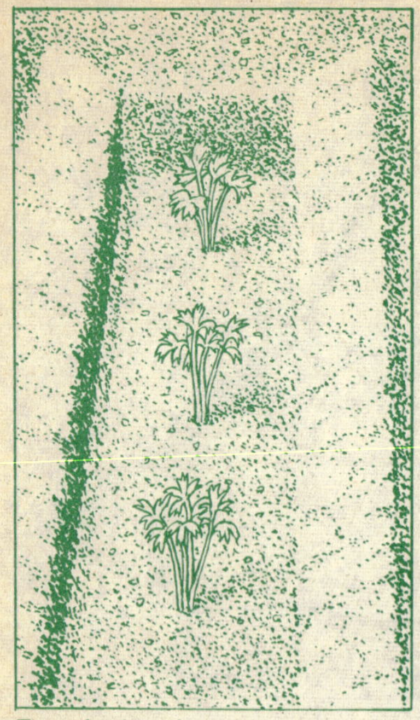

Transplanting celery

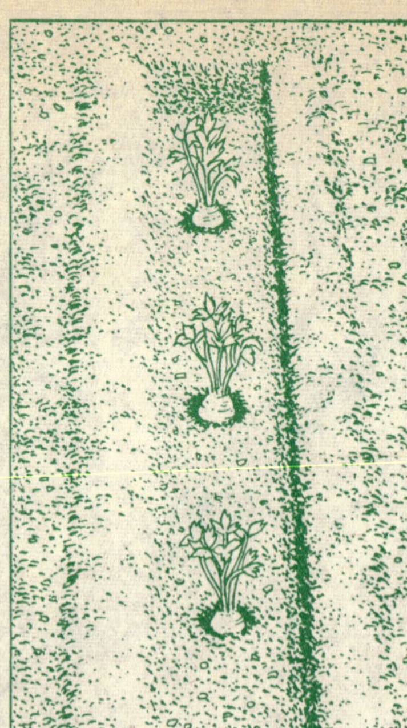

Transplanting celeriac

need to dig a trench, simply plant it about 23cm (9in) apart each way in holes, the depth of the hole being determined by the previous growing depth of the celery. The 'self-blanching' process is due to the fact that the plants are placed close together, thereby excluding the light, and of course it is the light which makes the plant turn green. Self-blanching celery is a crop which therefore needs a fairly large area to be successful.

A single row of the crop will simply not blanch itself. It needs the proximity of neighbours on all sides to blot out the light, and therefore produce the 'blanch'. It needs a lot of room so it has not been used on the kitchen-garden plot, although there is nothing to stop you planting some if you have sufficient space.

Transplanting celeriac

Celeriac is another plant with a slightly different planting out technique. Draw a flat bottomed drill about 5cm (2in) deep. Plant as celery, the flat drill allowing the bulbous growth to develop easily.

Celery, leeks and celeriac will all benefit from plenty of water. They should not be teased with water; plentiful but infrequent is the rule.

Transplanting courgettes, marrows and cucumbers

Courgettes, marrows and cucumbers can also be transplanted from seeds grown under glass, or from plants bought from the local nurseryman, or sown as described on page 19. To transplant, make a circular hole in the ground, fill the hole with water and allow the water to soak away. The plant should then be loosened from its pot by tapping on the pot. Having thus removed the plant, place it in the now moist hole and firm down the earth.

Pumpkins

A fruit grown in the same way as the marrow and courgette is the pumpkin. It may sound a bit exotic, but nevertheless it grows very easily in most parts of Britain, and pumpkins certainly look impressive. The *Kitchen Garden* pumpkin plant produced one fruit weighing in at 25kg (55lbs).

Growing tomatoes

Another fruit to look at now is the tomato. For a plot as small as the kitchen-garden plot it will probably be more economical to buy healthy plants from the nursery and transplant out in June. Tomatoes will thrive best on good humus. If you didn't put any humus into the ground on your first digging then do so now before you plant out your tomatoes. The plants purchased from the nursery should be about 25cm (10in) tall with a stem about the thickness of a pencil. The leaves should be a good rich green, and the whole thing should have the appearance of a good sturdy plant. Don't plant the tomatoes straight out into the plot; they need what is known as hardening off. Remember they have been in a warm cosy greenhouse and it could come as a nasty shock to them to be pushed out into the cold. The answer is to put them out by degrees. Put them out by day in their pots and bring them in at night, just for a few days before planting.

Give the ground a good soaking before planting your tomatoes and put stakes in the ground now—later they may damage your roots. Make a hole about the same size as the pot from which you are taking the tomatoes. Tap the pot to free the roots and put the plant in the hole and press down, not too close to the stem. Thoroughly water again and then leave for two weeks. Tomatoes thrive best on a lot of water, but do not water too frequently. So the rule is just the opposite of 'little and often': it is 'a lot and seldom'.

Tie the plant to the stake taking the lowest strong leaf, and tie underneath it, the leaves being the strongest part, and then twice round the cane. Do *not* tie under the truss or it may easily break off. Raffia is best for tying because it is less likely to cut into the stem as the plant grows.

Never tie below a truss

You will now be able to look out on your kitchen garden with great pride and a considerable sense of achievement. You should have over twenty crops in your plot. It has taken some hard work but as you see the seedlings coming through you will realise how worthwhile the hard work was.

5 Staking, feeding and spraying

One of the most important aspects of vegetable growing is the after care, the staking and tying and feeding, the pest prevention and so on, in fact doing all you can to keep your plants strong and healthy.

Staking

Staking can be done in various ways, but whichever method you use, the stakes should be put into the ground before you sow the seeds. In this way you will not run the risk of damaging roots and you will have a useful marking guide giving the position of the plants before they come up. These stakes must be firmly put into the ground, and driven in hard with the mallet. Later in the season the stakes will have to bear a considerable weight, particularly in the case of beans, and there could also be a good deal of wind—so hammer them in well.

Runner beans

Runner bean stakes should be placed in pairs about 60cm (2ft) apart with

about 125cm (4ft) between the pairs. A stake is leaned towards its partner and crossed at the top. A further pole is put across the pairs of sticks at the point where they cross, to form a ridge like on a ridge tent. The poles are then securely tied at the top to make a strong rigid construction.
Another method of staking is the wigwam construction. Take eleven stakes, each 250cm (8ft) long, and space them 40-45cm (15-18in) apart around a circle about 155cm (4ft) in diameter. The poles are then made to lean inwards to meet at the top and again securely tied off.

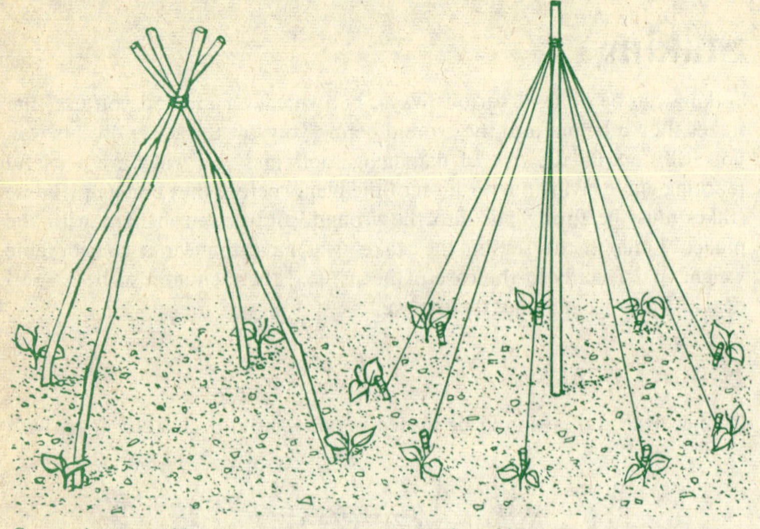

Some wigwam staking *Maypole staking*

A third, simpler and indeed cheaper method is called the 'maypole'. Drive a pole about 250cm (8ft) long into the ground vertically for 60cm (2ft) of its length, leaving 180cm (6ft) above ground. Tie strings to the top of the pole, and bring the strings out to the edge of an imaginary 125cm (4ft) diameter circle, with the pole at the centre. Fix these strings into the ground with hooks like tent pegs. Another advantage of this method, in addition to the low cost, is that it is possible to cultivate a fast maturing crop like lettuce in the middle of the circle, before the beans grow up the strings.

Peas

Peas also benefit from a limited amount of staking. Use twiggy sticks, about 90cm (3ft) high, and remember to put them in the ground before sowing the peas. Staking peas keeps them from trailing on the ground, thus preventing mould. It also helps them to ripen by holding them upwards to the sun.

Potatoes

Potatoes are not normally staked, but with the limited space on the plot, it may prove necessary. Potato greenery is liable to spread itself and fall over nearby crops, causing the neighbouring plants to fight their way through the potato leaves and in doing so, to become leggy. Four short canes about 120cm (4ft) long should be placed one at each corner of the potato plot. Tie raffia round the perimeter of the potato plot from cane to cane. As the potato top growth increases, tie another line above the first, and put in more canes if necessary.

While on the subject of potatoes, one must also remember to earth them up round the roots, building the earth up to cover the tubers, otherwise they will go green. Unlike the celery mentioned in the previous chapter, potatoes are not self-blanching.

Tomatoes

The stakes for the tomatoes should have been put down before transplanting the tomatoes to avoid damaging their roots (see page 29). The tomato plant should be put as near to the stake as possible. Tying should continue as the plant grows, and as page 30, the tie goes under a leaf, and not under a truss.

Growing tomatoes

Remove the side shoots from the tomato, as only the main stem is needed to provide sufficient tomatoes for all your needs. The side shoots appear between the leaves and the main stem and they should be pinched out with the thumb and forefinger, leaving the leaf intact. The top of the plant should be similarly pinched out when four trusses are formed. This prevents the plant from producing more trusses, too late to ripen in the open.

To encourage pollination and the production of fruit you could try damping over. Take a watering can with a fine rose and flick it upwards over the plant. This will allow a gentle spray of water to dampen the flowers. This should be done on a warm day, either early morning or evening, not in the heat of the day. As the trusses become well set and the fruit starts to form, then the watering should be increased. Do not tease the plants with water—give them a good soaking, and leave them for a week.

Plant feeding

The feeding of plants is not absolutely essential. Provided you have prepared your soil well at the beginning of the season, the plants should get all they need from the soil. It may be necessary to feed some crops, though, if they don't appear to be doing particularly well. Plants should certainly not be overfed. A top dressing of fertiliser should be all that is needed. Ideally the fertiliser should be applied when the ground is moist, as feeding in the middle of a dry spell can cause scorching of the roots. As for quantity, a handful of fertilizer scattered along either side of a 150cm (5ft) row should be all that is necessary. Hoe the fertiliser in with the dutch hoe.

Control of pests

Most plants will suffer from pests of one sort or another, but thanks to the use of modern chemicals, these pests are far more easily controlled than they were in Grandad's day, when black fly could do as much damage as Moses's plague of locusts. There are two main types of chemical spray in use. The first is sprayed directly on to the infestation in question. This has the same effect as squirting fly spray on to a fly, attacking the cause of the trouble at the moment it occurs.

The second type is called a systemic. With this type the chemical is absorbed by the foliage, and is therefore much slower acting. It can however be most useful since few people have the time to rush into the garden armed with a spray every time a greenfly is spotted. The systemic type of spray is also extremely useful for spraying plants before you go on holiday; slower acting they are also longer lasting.

Always spray below the leaves as well as above, particularly with the first type of spray. This deals with insects hiding under the leaves.

Caution with sprays

Certain chemicals can be used on certain plants while being unsuitable for others. On a small plot this obviously presents problems. It is almost impossible to spray one crop on one small part of the plot, at the same time avoiding adjacent crops, so you must in some way isolate the crop you are spraying. This can be done by covering all neighbouring plants with plastic sheeting or old plastic sacks, while you spray the crop in need of treatment. Wash the plastic sacks after use, and of course wash your hands, mixing vessels, containers and sprayers very thoroughly. Finally, a most important word of warning. Keep *all* chemicals correctly and clearly labelled, keep their lids securely on and **keep them out of the reach of children.**

Spraying equipment

There are a number of types of sprayer available, from the simple hand-operated pump, like an oversized scent spray, to the sophisticated pressurised back-packs. The simpler variety is quite adequate for the kitchen garden and should not cost more than a pound. The chemicals for the spray should be a further £2. Three different types of spray should cover most pests and many diseases. These chemicals will make up to about 136 litres (30gal) of liquid and should last about two seasons.
If time is too limited to mix chemicals, you could use aerosols. They are more expensive than mix-it-yourself chemicals, but they are certainly effective.

When spraying, whether with hand spray, back pack or aerosol, remember:

1. Read the instructions carefully.
2. Pick a mild and, if possible, windless day (your neighbour may not want his crops sprayed).
3. Cover up all crops that you do not wish to be sprayed.
4. Wash everything thoroughly that has come into contact with the spray.
5. Label all chemicals and keep them out of reach of children.

A further point to remember when spraying is that an insect spray may not always differentiate between the insects you want to get rid of and those that are useful, so don't use your spray indiscriminately.

6 Catch cropping

Many gardeners under-use their plots. After taking out their early maturing crops they leave the ground fallow until the following year. Occasionally it is a good thing to allow the ground to lie fallow. The *Kitchen Garden* plot is small, therefore to make the most of it you will need to use the ground again for more crops before the end of the growing season. This method of extending the growing life of your plot is called catch cropping. Simply replant the areas left vacant by early maturing crops with other crops that will mature before the late autumn. You can of course replant lettuces and radishes, peas, broad beans, and possibly potatoes or you can take the opportunity to introduce some of the more unusual vegetables to your plot and maybe to your family as well.

By mid-July your plot could be looking a little bare. The broad beans will have gone, likewise most of the lettuces and radishes, the spring onions and the potatoes, and shortly the dwarf beans will be leaving an empty patch behind them. Don't leave the plot bare. Sow more seeds, and enjoy more vegetables before the season is over.

First you will need to remove all traces of the old crops, except in the case of beans and peas. Usually a thorough hoeing and raking is all the ground will need to make it ready for re-sowing. Where you are replanting after beans and peas remove only the haulm. They should be cut off at ground level, and never pulled up. Both peas and beans have small nodules on the roots which contain nitrogen and this of course is highly beneficial to the new seeds you are about to sow.

It is advisable when catch cropping, not to plant the same crop in the same place. Move the crops around, put the lettuce in the radish bed, the radish where the peas were and move the peas to the lettuce bed. Certain plants will draw off certain constituents in the soil, and by changing the plants around you will avoid the earth of your plot becoming deficient in any nutritional constituent through the over-planting of one particular crop. The carry-over of disease is also less likely if the crops are changed over, not only just where catch cropping is concerned, but also when considering next year's plot. Rotation of crops is a very important factor in small-scale as well as large-scale growing.

The seeds you put in for catch cropping will need to be the early maturing variety since you will have to harvest them fairly soon. To start with, some of the less well known varieties. See also charts pages 20-23.

Kohl rabi

This is a somewhat unusual vegetable from the Continent. It tastes like a cross between cabbage and turnip. You can sow it from April until mid-July, so it is a good candidate for catch cropping. It matures in about 10-12 weeks from sowing. Sow the seeds in 25mm (1in) deep drills, and if you are planning more than one drill keep them 38cm (15in) apart. Later you will need to thin the seedlings to 75mm (3in) spacing.

Golden beet

This is another fast maturing crop that can be sown as late as July. Sow in 1cm (½in) drills, keeping the drills 30cm (12in) apart. Use the method described on page 18 for ordinary beet—groups of three seeds at 5cm (2in) intervals.

Salsify and scorzonera

Both these exotic sounding plants can be sown later in the year. They have the advantage of being 'winter hardy', which means they can be left in the earth through the winter and harvested when required. Sow them in 25-40mm (1-1½in) drills, leaving 38cm (15in) between rows, and thin to about 20cm (8in) between plants.

Winter radish

These are normally sown in July for harvesting from October onwards, making a marvellous addition to winter salads. Sow them in 25mm (1in) drills, 30cm (12in) between the rows and thin the seedlings to 75mm (3in) between each plant. Don't forget the 4 per cent calomel dust to prevent root fly.

Cauliflower

Some varieties of cauliflower can be planted in July to produce heads in late autumn. Sow in a shallow drill—30cm (1ft) between rows if you are sowing more than one row—and then transplant the seedlings so that you have 60cm (2ft) between the plants each way.

Cabbage

Another useful catch crop. Some varieties will go right through the winter. Sow as described on page 18.

Spinach

A late sowing of spinach can be extremely useful especially if your perpetual type is suffering from constant cutting. A good variety here is Greenmarket. Sow in a shallow drill about 25mm (1in) deep, and thin to 15cm (6in) between individual plants.

Parsnips

The parsnip is a slow maturing crop, and will use up your plot for a lot of the year without any yield, but if you have the space it is well worth growing. A handy hint—if you do contemplate sowing parsnips, sow one or two radish in the row with them. Radish, being an early maturing crop, will come through quickly and mark the row for you.

There are of course other crops that can be sown as late as July for harvesting in the autumn. A browse through your seed catalogue will tell you which ones. Catch cropping is an important part of the vegetable growing year, so don't ignore it or you could be turning your back on some delicious late autumn crops.

7 Harvesting for storing

It is hard to imagine today that there was a time when you couldn't rely on the cold cabinet at the grocer's for vegetables that were 'fresh-frozen'. But even in the pre-freezer days there were still methods of storing vegetables, so that you could enjoy many of the late summer vegetables all winter long.

The garden will still produce some vegetables in one form or another throughout the winter. It is possible for instance to harvest up to six different crops in January. Nevertheless it is helpful to be able to supplement these winter vegetables with some of those grown in the autumn and then stored. Not only does this give you more variety, it also makes certain vegetables readily available at a time when they are at their most expensive in the shops.

What vegetables to store

Some root crops will store through the late autumn until well past Christmas, the carrots and beet certainly into early spring. The beetroot used in the *Kitchen Garden* had been stored by the method described on page 40 for a whole year and was still edible. The storing process for carrots is similar to that of most root vegetables, such as beet, winter radish, turnips and kohl rabi.

Good storing starts when you take the vegetables out of the ground. Great care should be taken to see that they are not bruised or damaged in any way. Any crop that has an imperfection should be used immediately in the current cooking pot and not stored for use in a future one.

The second golden rule for good storing is to lift your vegetables from the ground when the ground is as dry as possible. Wet muddy vegetables will not store.

What you will need

You will need a wooden box for storing the vegetables.One approximately 60×45×60cm (2ft×18in×2ft) should be adequate, but for larger storings use a larger box. If you intend to store only a few vegetables, then the box can easily be subdivided with a simple partition making it possible to store two vegetables in one container.

You will also need a storage medium, that is, the stuff into which you will be putting your vegetables to keep them over winter. It can be either peat or sand, but peat is recommended. It should be just moist enough to be really effective, not too wet, not too dry. To test for this condition, take

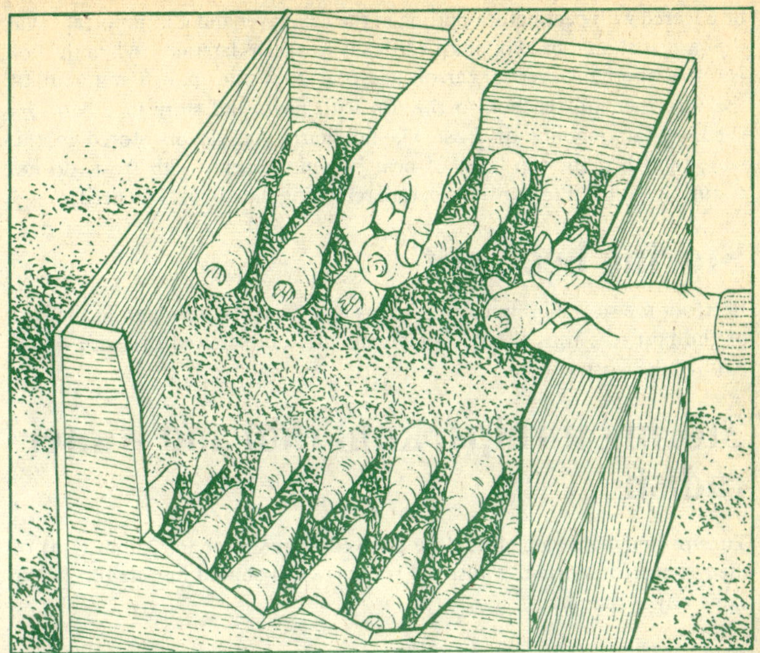

Storing carrots

a handful of peat and squeeze it in the hand—if moisture comes out between the fingers, then the peat is too wet. If the peat falls apart in the hand then it is too dry. It should just hold together. The peat needs to be in this just moist condition in order to protect the crops you will be storing. If the peat is too wet it could cause the vegetables to sprout or go mouldy, and if it is too dry it will cause them to shrivel up. The right consistency of peat will give you 'fresh' tasting vegetables all through the winter.

Place a 5cm (2in) layer of peat at the bottom of your storing box. If the slats at the bottom of your box are wide apart then you will need to line the bottom of the box with a layer of newspaper. Now you have a bed in which to store your carrots or other vegetables.

Storing carrots

First cut the green tops off the carrot, cutting them just where the green part joins the orange part. Lay them on the peat, point to point, with a small gap, say 5mm (¼in) between each one. It is important that the carrots should not be touching each other, nor in any danger of rolling towards each other and then touching. This layer of carrots will need to be covered with peat before putting in the next layer and it is important that the carrots should be kept apart vertically as well as laterally. When the box is full of alternate layers of carrots and peat, to within about 75mm (3in) of the top, then fill the box to the top with peat, and store in a cool dry shed. A fair amount of frost will not harm the carrots stored in this way, though if the top layer of peat becomes crackly with frost, do not disturb it. Allow it to thaw out before removing the carrots you need.

Storing beetroot

Beetroot are stored in much the same way as carrots, although in the case of the beet, the tops should not be cut off, but screwed off with a deft twist of the wrist.

Storing turnips, kohl rabi and winter radish

Turnips and kohl rabi should be allowed to retain about 25mm (1in) of their stems for storing, whilst the winter radish should lose its top in the same way as the carrot.

Storing parsnips

One other root vegetable, the parsnip, has even fewer storage problems. It can be left *in situ* in the earth all winter and harvested as required.

Storing beans

Beans are another vegetable that can be stored, particularly the Chevrier Vert or the climbing French bean—the variety called 'Earliest of All'. The Chevrier Vert is a particularly useful bean as it can be harvested in a very young state for immediate use, or left until an intermediate stage to give young green beans, or flageolets, when shelled. Finally, if they are left on the plant they will ripen and dry and when shelled they will produce the dry white bean, the haricot blanc, for the bean pot, the cassoulet, or as the basis of baked beans. The climbing French bean will also produce a small dry white bean for winter storage, which can be cooked in similar ways to the haricot blanc. Beans can be stored in a paper bag or in a glass jar with an air-tight lid.

Storing onions, shallots and garlic

Onions are an ideal plant for storing—ask anyone who has bought onions from the blue-bereted Breton onion sellers who cycle round the South of

England selling their wares. Onions are not difficult to store, but it is important that they should be in a good state for storing. The tops should be shrivelled and lying over on the ground, whilst the bulb should be a golden brown. Care should be exercised when lifting the onions, as with all other crops for storing. It is important that the onions should not be damaged and they should be eased out of the ground with a fork, taking care not to tear the roots. Ideally they should be left lying on the ground for a few days to dry out fully. If the weather does not permit them drying off outdoors then take them inside and allow them to dry off under cover. Remember that although onions may look hard and solid, they will bruise as easily as an apple, so handle them with care. Once the onions are dry they can be tied up by their foliage, either in long strings, like the Breton onions, or just taken in bunches of about eight, tying raffia round their foliage. Hang them up in the roof of the shed or some other cool dry place. It is possible to store them in an open shed as they are not affected by frost.

It is not essential to tie the onions into bunches. You can, if you prefer, cut the tops back to within 5cm (2in) of the bulb and then store them in a string bag or sack (*not a plastic sack*) and hang it up somewhere where the air can circulate. They can even be stored on slatted shelves, as long as there is the essential circulation of air.

Shallots and garlic can be stored in the same way.

Storing celeriac and leeks

Celeriac and leeks are two other vegetables which can be stored very satisfactorily over the winter months. Both vegetables can be stored out of doors in trenches which you will need to dig in December when clearing out the plot in preparation for the next year's sowing. Both vegetables can in fact be left in the ground, but by storing them in the trench system, less space is needed and the ground can be cleared at the same time. You will need a trench about 10-12cm (4-5in) deep for both celeriac and leek, but the trench for the celeriac will need to be slightly wider than that for the leek. Considering the rather bulbous nature of the celeriac, it will also need to be flat-bottomed trench, while the leek will only need a v-shaped trench. Having prepared a home in which to store your vegetables, you will need to prepare the vegetables themselves. The celeriac will need to lose much of its greenery. Slowly and carefully take off the greenery at the top of the bulb. As you remove the greenery, take care not to remove the leafy stems right in the middle of the plant. This, with the bulbous root, is what you will need to store over the winter. Put the celeriac root into the flat-bottomed trench to the same level as it was in its original growing

Storing celeriac and leeks

place, thus allowing the remaining bit of greenery to show through the surface. You will now know where to find your celeriac. Put the celeriac into the ground about 25mm (1in) apart from each other. Although you are storing them in the garden, they are taking up much less space than they did when they were growing.

Leeks will need the green part cutting down to about 10cm (4in) above the white edible part. If this is not done, the leaves could well fall over and become mildewed on the ground. This deterioration then passes down the leaves to the roots of the plant, making it inedible. Having cut off the top green, place the leeks in their trench to the same depth as they were before you took them out, leaving about 1cm (½in) between each leek. Cover up your vegetables with the earth, making sure of course that their green bits show above the surface.

Storing herbs

Parsley, thyme and sage will continue in the open for some time into the winter, but if you want to store them you will need dry weather and a crop in prime condition. Cut yourself an easily manageable bunch, tie it round the stalks and hang it up to dry in a cool, dry, airy place. When the leaves become brittle rub them down between your forefinger and thumb, and store them in a glass jar with an airtight lid. A screw top jam jar is ideal, but paper bags will do in a really dry place.

Storing marrows and pumpkins

Marrows and pumpkins, the simplest of all crops to store. You can keep them almost anywhere that is dry, and they will last half the winter without any fear of deterioration. No special beds or trenches, no preparation, just keep them somewhere dry and airy and they will obligingly be there and ready whenever you want to use them.

Short storing period: cauliflowers

Brassicas are not generally considered to be good subjects for storing, particularly when it is possible to grow sprouts, kale, spinach and cabbage all the winter. Cauliflowers can be kept for a short period of time, say about a month. If they are likely to be damaged by early frost, lift them, complete with root, and lay on a cold floor. To prolong their life even longer, wrap the root with wet soggy newspaper, well tied round with string, and hang upside down from the roof of the shed or any other cool place.

Well, there it is. An enthusiastic amateur's guide for, I hope, thousands more would-be enthusiastic amateurs. A guide to one small area of gardening that is remarkably uncomplicated. It's really not difficult at all and once you have had a try, your confidence will grow.

But the most important thing of all is that gardening should be a pleasure. This small book does not contain a chapter entitled 'Standing and Looking'. If it did, it would contain just two words—'Do so'. Stand and look when the soil is dug and hoed and raked and ready for planting. Stand and look when the seeds are just showing through. Stand and look when the peas and beans come into flower, when the first truss is set on the tomatoes, and when those luscious large yellow flowers bloom on the marrow plants. Without expending any energy, without bending your back, without even flexing a muscle just stand and look and the sense of pleasure and satisfaction is immense. That's gardening.

VEGETABLE RECIPES
CLAIRE RAYNER

8 Making the most of your vegetables

Vegetables provide just about every food element we need—proteins and fats and carbohydrates, vitamins, mineral salts, roughage—you name it, and a gardener can grow it!

Unfortunately a lot of good gardening effort is ruined by indifferent cooking—so before going on to discuss recipes for individual vegetables, let me offer a few basic rules. And let me start by saying the best advice any would-be vegetable cook can be given is the same that Victorian Mammas used to give their daughters when they launched them on to a social life: 'Don't—but if you must, do it as little as possible'. In other words, vegetables, especially freshly gathered home-grown ones, are best eaten raw. Not only do they taste better and look prettier—they put their stores of vitamins and minerals into you, instead of down the drain with the cooking water. And there are very few vegetables that you can't eat raw.

Cooking vegetables

Those that must be cooked, however—potatoes, for example—can be kept at their best if they aren't over-prepared. Peeling is one of those kitchen jobs that waste food, time, and a cook's energy. The merest scrubbing is all most of 'em need (I use a large nailbrush) and then you can cook them in the oven or on the hob. I much prefer baking myself, and that *doesn't* mean that you always have to have potatoes, for example, baked in their jackets, good as that is, or roasted in a lot of fat. Use foil to cook your vegetables: slice them, and put them in the middle of a sheet of foil with seasoning, a little butter (or margarine), or no fat at all if you want to cut down the calories, parcel it up firmly, but not too tightly (you need room inside the parcel to let the steam circulate!) and bake in a moderately hot oven, Regulo 5 or 6, 190°C-205°C, 375°F-400°F. Depending on the thickness of the pieces, it takes from half an hour to an hour. This method can give you baked mashed potatoes—or baked scalloped potatoes—or baked steamed potatoes—and it saves fuel, too, if you have the oven on anyway. Use it also for carrots or turnips or marrows, especially baby marrows (courgettes) or any vegetable you'd usually boil.

When it comes to leafy vegetables, however, you can't so easily use the oven. But you still don't have to boil the living daylights out of the poor things, in the classic boarding-house landlady's fashion. Shred finely first of all, or if it's cauliflower or such, break in small flowerets. Then, *steam them*. I use a colander over a saucepan—no need to buy fancy special pots—and cook just till they're bite tender. Spinach you can cook in butter (cabbage too, come to that) with a minimum of water. I give delicious recipes for spinach on pages 50 and 57.

Right now, back to basic principles. And let's consider the worst method of all, when it comes to calories, but one of the nicest when it comes to taste. Deep frying. Marvellous for courgettes and onions as well as the obvious potatoes. Slice the vegetables into bite-sized pieces, wash, make sure they're well dried after washing and then dip first in seasoned beaten egg, then in seasoned flour or coarse ground oatmeal, or even crushed cornflakes if that's what you fancy. Shake off the excess, and fry in a deep pan when the oil is smoking hot. Drain well, and eat it right away. Deep-fried foods sag a bit if they're kept waiting.

Seasoning

Having mentioned seasoning, what about the best seasonings for vegetables? There's more to life than salt and pepper—even coarse sea salt and freshly ground black pepper, both of which I swear by because the taste difference is indescribable. Until you've tried them, you won't know how boring or dreary ordinary table salt and ground white pepper are! You also need herbs and spices. Some people are very hide-bound about what goes with what, always following rules. But you don't have to! Cloves are great with apple pie, but try them with puréed turnips too. Nutmeg goes naturally on top of a milk pudding, and even more naturally with potatoes. Try baby ones fried whole in butter and sprinkled with nutmeg. About nine million calories, but the taste is fabulous.

When it comes to herbs, freshly chopped ones are always best. To add to those from the garden, I grow a lot on the kitchen windowsill. You can never grow too many; if you get a heavy crop, keep extra supplies in polythene bags in the fridge (or freeze whole leaves). Crush them before you open the bag, and you've got it ready chopped. I use this method for mint, and basil, and marjoram, and thyme. I know I said there are no rules about what goes with what, but let me just offer a few suggestions: try basil with any tomato dish; thyme in a salad dressing; and marjoram with cabbage. They make an enormous taste difference!

Do remember that all vegetables can be enhanced with clever additions. Like almonds fried in butter and mixed with cauliflower or broccoli. And butter-fried breadcrumbs tossed over puréed turnips or potatoes, and grilled for a final fancy touch.

Cooking spaghetti marrow

In my first chapter, as well as talking about these basic principles I would like to tell you how to cook a most exciting and exotic vegetable—*spaghetti marrow.* I always knew the stuff grew on trees, if only the truth were known. Easy to cook—just weight it with a glass plate in a big saucepan, so it stays well under water, and boil it whole for around forty-five minutes to an hour, depending on size. It will feel springy when it's done. Then the tricky part. Slit it in half longways and scoop out the seeds. Then scrape it out of its shell, and fork it up, as though you were scratching it. It will go into long green strings—that's how it got its name. I like it served with a rich tomato sauce and a sprinkle of parmesan cheese, or you can serve it hot with melted butter and salt and pepper, or with chopped bacon, or chopped fish, or cold with an oil-and-vinegar dressing. Any way you'd serve real spaghetti in fact! Very filling, not as fattening as real spaghetti, and it tastes great.

Tomato sauce for vegetable spaghetti

SERVES 6

metric	*imperial*
25g. butter *or* 1 tbsp. oil	**1oz. butter *or* 1 tbsp. oil**
1 large chopped onion	**1 large chopped onion**
1.25-1.5Kg. fresh, skinned chopped tomatoes	**3lbs. fresh, skinned, chopped tomatoes**
Brown sugar; basil; sage; salt; pepper (to taste)	**Brown sugar; basil; sage; salt; pepper (to taste)**

Melt butter in oil, cook onion till transparent, add tomatoes and seasoning. Simmer till rich and dark in colour—one hour at least.

9 Onions, peas and beans recipes

In this chapter, I go on to two of the most staple vegetables. Onions and pulses—the edible seeds of leguminous plants according to the dictionary—peas 'n' beans according to us.

Onions are, of all vegetables, the most indispensable. There are very few savoury dishes that don't need them and they're also a splendid vegetable in their own right. They're high in carbohydrate and vitamin C and mineral salts, especially potassium and calcium—and they have been said to give protection against heart disease! That's not proven, though, but their taste is. So try them baked in their jackets, just like potatoes and eaten with butter, or stuffed with a herby breadcrumb mixture, or of course deep fried in rings, or simply braised.

Peas and beans are also versatile—put them in soups and stews and salads—and are also an important source of protein, mineral salts, and some vitamins. There are areas of the world where people live on virtually nothing else. Most people cook only the peas and throw away the pods. But do remember you *can* cook the pods, especially when they're young, with a ham bone, a few root vegetables for added flavour (onions and turnips) and seasoning. Cook for a couple of hours, take out the bone, and put the rest into a blender, and you've got a great pea soup.

Peas

Anyway, here is a method with peas that really is delectable—even better than the classic peas-with-mint. Take small onions, the cocktail size (and if you peel them *under water* your eyes won't leak—a useful tip, that) and toss them in a butter and oil mixture at the bottom of a heavy pan. I always use a little oil with butter to prevent the butter burning. Olive oil when I can afford it, ordinary vegetable oil when I can't. They'll go a little transparent after a few minutes. Then add a little sugar, a bunch of fresh parsely (tied in a little bundle so that you can remove it easily afterwards), salt and pepper, a little hot water and the shelled peas. Then put a small quartered lettuce on top, lid it tightly and leave it all to cook on a low heat for about a quarter of an hour. Longer if the peas are a bit elderly. Serve it with the quartered lettuce arranged round the edge, and add a little more melted butter. Smashing!

Beans

To cook runner beans, you need only to steam them and serve them tossed in a simple oil dressing—olive oil seasoned with sea salt, black pepper and garlic. And they're smashing cold, too, with a vinaigrette sauce. Broad beans can be treated the same way—though they're best eaten raw when they're very small—or tossed in a light bechamel sauce when they're getting more mature.

Mange-tout

Beans aren't the only things you can cook in their pods. A special vegetable is *mange-tout* (eat-all). The best way with these is to string them by pinching off the tip and pulling lightly, so that the fibrous edging of the pod comes away. Then cook them for a bare ten minutes or so in melted butter. I use so much butter I ought to get the Cow-of-the-Year award, but honestly, it's unbeatable for vegetables, whatever it does to your hips. However, you can use margarine, and a low-cholesterol one at that if you are worried about the amount of this substance you get in your diet. Cholesterol, by the way, is the form of fat you get from animal foods (meat, butter, milk, eggs) and some doctors say a low-cholesterol diet, using only fats derived from vegetables (such as corn or olive oil) is better for your health because cholesterol has been implicated in some forms of heart and artery disease. However, not all doctors agree on this point, and fear sugar far more than fat! But, if you want to play safe, go for vegetable oils and low-cholesterol margarine.

Now back to my *mange-tout*! The French always cook these with a sprig of savory, but I'm not keen on it—I think this herb has a bit too powerful a taste for their delicate flavour. You can serve these as an accompanying vegetable, but I think you get more out of them eating them as a separate course in their own right. That's true of most vegetables, as a matter of fact—we in Britain tend to see vegetables always as bridesmaids, meant to hide in the shadow of the meat or fish. But like bridesmaids, get 'em on their own, and they can be much more interesting.

Onion bread

To finish off, I offer a recipe of my own—onion bread. I use a basic pizza dough, spread it into a 1cm (½in) thick square in a shallow oven dish and spread it with onions which have been fried till transparent in butter-and-oil (or if you prefer the flavour, beef or pork dripping) and seasoned with whatever you fancy. I like basil, or sage, or marjoram—my mood varies—and always, of course, sea salt and black pepper. Then you cook it flat, and serve cut into squares, or roll the whole thing into a sausage or swiss roll shape, cut it into rounds, leave them to prove—that is, rise again—and bake in a hot oven (Regulo 7, 215°C, or 425°F) till they're brown and crisp. It takes about half an hour. You can sprinkle the tops with a few more fried onions mixed with poppy seed, for a crunchy finish. Eat 'em hot! Great for family suppers, smashing for parties.

Bechamel sauce for broad beans

SERVES 6-8

metric	*imperial*
90g. butter	**3ozs. butter**
4 heaped tbsp. plain flour	**4 heaped tbsp. plain flour**
9dl. milk	**1½pt. milk**
Nutmeg; salt; pepper	**Nutmeg; salt; pepper**

Melt the butter, stir in flour to a thick paste, then add seasoned milk *slowly*, stirring all the time. Cook till thick and smooth and the flour is completely cooked. Pour over 90g. (2lb.) cooked shelled beans.

Pizza dough for onion bread *MAKES 1 DOZEN SLICES*

metric	*imperial*
45g. fresh yeast *or* 12-15g. dried yeast	**1oz. fresh yeast *or* ½oz. dried yeast**
A little sugar	**A little sugar**
Warm water	**Warm water**
450g. plain flour and pinch of salt	**1lb. plain flour and pinch of salt**

Liquefy yeast in sugar and water. Add to flour to make a well-kneaded dough. Leave to rise to double its bulk (about ½-1 hour).

10 Cabbage, spinach, sprouts recipes

Now for brassicas—the leafy vegetables. The studio kitchen looked like Kew Gardens, it was all so green! (I thought I was being grown something to wear when I first heard about them. Instead I showed how to cook them.)

Cabbage

The cabbage family have been sadly abused in this country in the past. Shades of all those horrible school dinners with their overboiled wet cottonwool greens. Ugh! So, first of all, please *don't* boil cabbage in slabs. It's a sin and a crime, really it is. It should be shredded very finely and steamed in a colander-in-a-saucepan steamer. Or better still, try it this way.

Chopped onion is cooked to a gloss in butter and a hint of oil. Now add a shredded cabbage mixed with a teaspoonful or so of caraway seeds and a hint of salt. You have to experiment to get the taste that suits you. Stir it vigorously over low heat. After a little while the natural cabbage juices will start to run. That's the trick—with cabbage and spinach—and once you've got it, you won't need any water. You can add a little, if you feel you must, and then it's safe to cover it tightly and leave it over a miniscule heat to cook. Stir it occasionally to prevent it sticking, and when it's crisp and *al dente*—rather like spaghetti—add the juice of half a lemon. If you like you can then sprinkle the cabbage with flour and stir till that's cooked—another five minutes—to make a sort of thin sauce, but do remember that adds calories! This is especially good with any rich pork dish.

Spinach

For spinach, use the same method, but leave out the caraway seeds. Instead use nibbed almonds, if you want to be really exotic. Or use a chopped onion and then serve the spinach with a poached egg in the middle—the classic Florentine dish. A marvellous lunch!

Sprouts

What about sprouts? They can be steamed of course—and never ever commit the dreadful sin of putting bicarbonate of soda in them to hold the colour. They may look greener but it ruins the taste. Also, cut a deep cross shape at the bottom of each sprout before steaming it—this helps to make sure the stalk is adequately cooked through. Then try one of the additions I mentioned in Chapter 8, such as blanched shredded almonds fried in butter until they're brown. Very luscious, I promise you. You can do the same with broccoli and cauliflower. You can also use chopped chestnuts with sprouts—that's a classic Christmas thing of course.

Chinese cabbage chop suey

But let's get on to the really special dish—Chinese cabbage. It's an elegant looking vegetable, so much tidier than our sprawling British versions! It's great eaten raw like lettuce—and try cooking it Chinese style.

But first of all, do your own bit of growing—bean sprouts! Moong beans (get them in an Indian or Chinese grocers) are soaked for several hours (overnight will do) and then wrapped in a towel, and put in the airing cupboard in the dark for two or three days. Shake the towel occasionally to prevent them taking root on it. Wash and eat. They're lovely raw. But in this recipe, I cook them!

Chop a Chinese cabbage very small (really bite-sized) and take a big flat pan you can use on top of the cooker. You will need a really *big* shallow one. Put in it a little oil, and toss in the cabbage, and cook it fast, stirring like mad all the time. They call it stir-frying, the Chinese. A very accurate language. Now, you can also add other vegetables—baby corn cobs, whole, and sliced mushrooms, and of course the bean sprouts, and cashew nuts, and finely sliced ginger root which adds an authentic Chinese flavour. If you want to be extra exotic, a little sliced pineapple is fun too. It can be eaten at this stage, all hot and crunchy. Lovely! But let's go on a bit. Add a little stock—chicken stock is excellent—with soy sauce to taste (soy sauce really *is* the authentic Chinese flavour), then bring that to the boil, and cook for three minutes, still stirring. Then add cornflour well mixed with butter. This will thicken it all a little. And there you are—your actual Chinese vegetable chop suey! Eat it alone, as a starter, or with plain grilled fish or meat.

Chinese cabbage chop suey

metric	*imperial*
1 chopped Chinese cabbage	**1 chopped Chinese cabbage**
Oil	**Oil**
Beansprouts; baby corn cobs; sliced mushrooms; sliced green ginger root	**Beansprouts; baby corn cobs; sliced mushrooms; sliced green ginger root**
3dl. of chicken stock	**½pt. of chicken stock**
Soy sauce (to taste)	**Soy sauce (to taste)**
25g. cornflour } mixed	**1oz. cornflour } mixed**
55g. butter }	**2oz. butter }**

Add cabbage to hot oil in a big pan, stir well, then add all other vegetables. Pour in stock and soy sauce, thicken with cornflour and butter mixture, serve at once. It takes 10 minutes to prepare.

Cabbage with caraway

Butter and oil
1 chopped onion
1 chopped cabbage
Salt to taste
Caraway seeds (about 1 teaspoon)
Juice of half a lemon

Melt butter and oil, add onion, then cabbage and seasoning. Stir well till juices begin to run. Cook *very* slowly till just crisp, add lemon juice and serve.

Takes 5-10 minutes to prepare. (You can add a sprinkling of flour to thicken the juices if you like. Cook for another five minutes in this case.)

11 Cucumber, courgette, marrow, pumpkin recipes

Fruity, now, because cucumbers and courgettes and marrows, which are this chapter's ingredients, are all fruits. So is the special—pumpkin.

Let's be honest and say that these vegetables aren't exactly your all-time taste sensations. Bland is the word for them, apart perhaps from cucumber, which has a very distinctive flavour when it's fresh. Do remember that there are some flavours that really bring out the best in cucumber—anise and caraway are examples. So does mint.

Cucumbers

Try this Greek way with cucumber. Chop it, sprinkle it with freshly chopped mint, and season with salt and garlic and then dress with plain yoghurt. I make my own yoghurt, from irradiated milk. To 1 carton of plain yoghurt (and try to get a good Balkan type from a health food shop, though you can use the ordinary kind) add 1.25 litres (2 pints) of UHF treated milk, or sterilised, or home-pasteurised milk (brought to the boil, then left to cool). Put in a warm place until set—about 12 hours in my airing cupboard but it may be more or less in yours. All cupboards vary! After making the first batch, you use the last dollop of it to start the next. Add fruit, nuts, honey and so on for sweet dishes, or use it plain for savoury ones.
Cucumber is also good cooked. Try this very delicate method. You need diced cucumber, chopped skinned tomatoes, chopped onions, seasoning. Sweat the onions in butter and oil until they're transparent. (So many recipes start like that, don't they?) Then add the cucumber and seasoning and go on cooking till the cucumber also goes transparent. Takes about fifteen minutes, maybe more. Then add the chopped tomatoes and go on cooking till the tomato sort of melts. Keep the lid on tightly. This stage takes about fifteen minutes more. Finally add chopped parsley and a little more seasoning. Another moment or two to heat the herbs through and it's ready to serve. Great with fish, or any delicate meat, such as veal or chicken.

Marrows and courgettes

You can do the same thing with marrow, by the way, as long as you cut it small enough, and it's just as good, only different.

Now we've got on to the subject of marrows, what about courgettes, which are baby marrows? I've already discussed deep frying them, but how about this? Slit largish ones lengthways, and scoop out the innards, leaving a sort of rowboat shaped shell. (It's easiest with a small spoon.) Chop the flesh, and mix it with breadcrumbs, chopped anchovy fillets, chopped black olives, oregano and parsley to taste, plus garlic, the quantities depending on the size of the courgettes and your own taste. Use wine to moisten the breadcrumbs if you're feeling exotic, a little milk if not. Season with black pepper, pile the mixture back into the shells, and arrange on an oven dish. Trickle over them some olive oil, and bake in a moderate oven, Regulo 5, 190°C, 375°F, for about twenty minutes. These can be eaten hot or cold.

Pumpkins

Now to the fancy stuff—pumpkin. They can be terrifyingly huge vegetables. I quailed when I first saw a really big one. No wonder it was pumpkin the fairy godmother used to make Cinderella's coach. Nothing else is big enough! But fear not, they are tameable creatures. You can either scoop out the flesh, chop it, cook it in the minimum of water and then drain it well to make a purée, or shove it in the oven and bake it till the flesh is soft enough to be scooped out. That's the way Americans do it, I'm told—well, they must have not only enormous ovens but asbestos fingers if they do, because when I did it, I got thoroughly scorched! Whatever method you use, once you've got the purée, the fun begins.

Pumpkin pie

For me, there's nothing to beat an old-fashioned pumpkin pie. Thanksgiving style! One pie shell. I use a *pâte brisée*—the classic imperial 4-2-1 mixture, 4 ozs. flour, 2 ozs. butter, and 1 egg or now in metric 125g. flour, 60g. butter and 1 egg. It makes a nice crisp base. Now, the filling. To 4.5dl. (¾pt.) purée, add 180g. (6 oz.) brown sugar, or 90g. (3 oz.) sugar and the same of black treacle (molasses). I had some left over from the sugar famine (amazing what we bought then, wasn't it?) and it does add a special taste and rich darkness. Then half a teaspoon of cinnamon, a good pinch of nutmeg, and cloves, and ginger (you'll have to do this according to taste. I like it spicy) and the juice and peel of half a lemon, and the same of half an orange. (I keep grated peel frozen all the time—very handy.) Next, add the yolks of 3 eggs, blending well, and thin it down with a little milk—no more than 1.5dl. (¼pt.). Finally, blend in the stiffly beaten egg whites lightly and expeditiously. Pile the mixture into the shell, and sprinkle the top with some more peel and spices. Bake for about half an hour or so in a hot oven (Regulo 7, 215°C, 425°F) or till it's brown. It can be eaten hot or cold and goes well with ice cream! You can, by the way, make a soufflé only by adding extra milk to the mixture, and shoving it in a soufflé dish instead of a pie shell and, of course, definitely eat it hot and fluffy, fresh from the oven.

Cucumber with tomatoes

SERVES 6

1 large chopped onion
Butter and oil
Seasoning
1 peeled chopped cucumber
5 large skinned and peeled chopped tomatoes
Parsley

Cook onion in seasoned butter and oil, add cucumber, cook for 15 minutes in a lidded pan. Add tomatoes and cook in a tightly lidded pan on a low heat for another 15 minutes, adjust seasoning. Sprinkle with parsley. Serve hot or cold.

Stuffed courgettes

SERVES 4

metric	*imperial*
4 courgettes (largish)	**4 courgettes (largish)**
80g. breadcrumbs	**3ozs. breadcrumbs**
6-8 anchovy fillets chopped	**6-8 anchovy fillets chopped**
6-8 black olives chopped	**6-8 black olives chopped**
Oregano, Parsley, Garlic } to taste	**Oregano, Parsley, Garlic } to taste**
Wine or milk (to bind breadcrumbs)	**Wine or milk (to bind breadcrumbs)**
Black pepper	**Black pepper**
A little olive oil	**A little olive oil**

Mix all ingredients with chopped courgette flesh, moistened with wine or milk. Pile back into the shells. Bake in moderate oven Regulo 7, 190°C, 375°F for about 20 minutes.

Alternative stuffings for courgettes

Replace olives, anchovies and oregano with:

Tomatoes; chopped salami or ham; basil
Fried onions; grated cheese; mustard
Mushrooms; chopped chicken livers; paprika
Tomatoes; sardines; fennel
Minced cooked beef; cooked rice; thyme

Pumpkin pie

MAKES 8 GOOD SLICES

Pastry shell	*Pastry shell*
125g. flour	**4oz. flour**
60g. butter	**2oz. butter**
1 egg	**1 egg**
Filling	*Filling*
4.5dl. (drained) pumpkin purée	**¾pt. (drained) pumpkin purée**
180g. brown sugar	**6ozs. brown sugar**
½tsp. cinnamon	**½tsp. cinnamon**
Big pinch of nutmeg; cloves; ginger	**Big pinch of nutmeg; cloves; ginger**
½ lemon—juice and peel	**½ lemon—juice and peel**
½ orange—juice and peel	**½ orange—juice and peel**
3 eggs, separated	**3 eggs, separated**
1.5dl. milk	**¼pt. milk**

Mix purée with sugar, add spices, fruit rinds and juices, blend in egg yolks, a little milk, and finally well-beaten egg whites. Place in pastry shell, bake half an hour or so at Regulo 7, 215°C, 425°F.

12 Salads

There's no law that says you can't eat salads in the winter. There are some marvellous concoctions that are just as filling as steak and kidney pudding, and much richer in vitamins and minerals. Sometimes people suffer actual vitamin deprivation in wintertime because they don't think of eating raw vegetables.

So let me suggest one of my favourite winter salads. Shred raw cabbage—and the thinner the shreds the nicer. Add shredded carrot, very fine onion rings and a thinly sliced pepper or two if possible. Mix them well, and dress them with a handful of capers (or you could use chopped pickled onions). I use this concoction of my own to dress it. I mix paprika, finely grated parmesan cheese, garlic salt, a few dill seeds, celery seeds and poppy seeds (for crunchiness) and coarsely ground black pepper. These are tossed together in a polythene bag (which is the easiest mixing way I know) and then sprinkled lavishly over the salad vegetables. Then, I add a simple vinaigrette dressing—two parts of olive oil to one part of wine vinegar, and a spoonful of sugar, all shaken up together. Toss the salad, and let it stand for an hour or so in the fridge, and then repeat the spice sprinkling, just before you serve it. This goes very well with hot meat. Try it with gammon, or roast pork.

And what about hot salads? Try potatoes or celeriac boiled for half an hour or so, till they're cooked but still firm, cubed while they're hot, and then dressed with a classic mayonnaise which has been spiced with a dollop of mustard, lemon juice (or vinegar if you prefer it), pepper and garlic. You can pretty it up with chopped watercress. Serve it with frankfurters, or with grilled gammon steaks.

And another favourite hot salad of mine—freshly cooked whole beans (*chevrier vert* are good for this) to which you add chopped radishes and classic oil and vinegar dressing. And then—here's the fancy bit—crumbled crisply grilled bacon rashers are strewn over the top. This really does taste out of this world.

And since we've got to the bacon bits, why not make a raw salad out of a vegetable most people cook? Spinach. Tear washed dried spinach into neat pieces, add the same simple vinaigrette dressing—and then add crumbled Stilton cheese and bacon crumbs, and serve it right away, crisp and well dressed. By the way, eaten raw like this, spinach is rich in iron. If you cook it, the iron is no good to you any more because cooking changes it into a form in which it can't be used in the body. I always knew Popeye was as stupid as he looked!

Next, here are a few of my favourite salads of the more classic summer type. There is one that can be a main meal—Niçoise. Lots of crisp salad vegetables—lettuce, thinly sliced onion rings, radishes, pepper strips, watercress, cucumbers (whatever your own tame gardener brings in)—are arranged in a handsome bowl. Now, on top, arrange broken pieces of tuna fish, a handful of olives, quartered hard-boiled eggs, anchovy fillets, and finally, French dressing. There! Crusty French bread and a bottle of plonk, and you've got all heaven on a tray.

Also, try shredded kohl rabi with sliced radishes in a French dressing, which is a simple vinaigrette in which a spoonful of made French mustard has been shaken.

How about a simple green salad with a special dressing? This is my own version of a Caesar salad dressing, but without the raw egg which some people dislike. Into a screwtop jar put two parts of oil to one part of lemon juice (or wine vinegar), parmesan cheese, chopped spring onion, a crushed clove or two of garlic, dry mustard, sea salt and black pepper, and a little Worcester sauce. As ever, quantities to individual taste! Shake this together like mad and pour it over the salad. Mix well, and last of all strew it with croûtons—cubes of bread which have been fried to a golden-brown in garlicky oil-and-butter and drained well. Eat at once—it goes soggy if it's left standing.

And while we're on to dressings—try Hungarian cucumber salad. It's the dressing that makes it. Paper thin slices of cucumber are lightly salted and then marinated in a sweet and sour mixture made by pouring 3dl. (½pt.) of boiling water over 1 tablespoon of brown sugar, and then adding 3dl. (½pt.) of wine vinegar. Let the salad stand in the fridge for about an hour before eating it.

Or try this version: thin slices of cucumber (cut 'em longways for a change) and thin onion rings are sprinkled with caraway seeds or anise or dill, or curry powder for a really exotic touch (I've used them all at different times), and then dressed with sour cream. Chill well before serving.

Soured cream is great for any number of salad dressing and so is plain yoghurt—which is less fattening of course! You can add any number of different spices (paprika, crushed coriander seeds, garlic). Do remember that different herbs make an enormous difference to the basic oil and vinegar dressing—I use mint, or tarragon, or lemon thyme, or basil or chervil—just take your pick!

Hot salad

Potatoes or celeriac
Mayonnaise with mustard
Chopped watercress
Frankfurters

Winter salad

Shredded cabbage
Shredded carrots
Fine onion rings
Sliced pepper
Handful capers

Dress with:

Paprika
Grated parmesan cheese
Garlic salt
Dill seeds
Celery seeds
Poppy seeds
Ground black pepper
Vinaigrette dressing

Chevrier vert salad

Chevrier vert beans
Chopped radishes
Oil and vinegar dressing
Bacon rashers fried and crumbled

Spinach salad

Spinach
Vinaigrette dressing
While Stilton cheese, crumbled
Bacon crumbs

Niçoise

Lettuce
Sliced onion rings
Radishes
Pepper strips
Watercress
Cucumbers
Tuna fish
Olives
Hard boiled eggs, quartered
Anchovy fillets
French dressing

Kohl rabi dressing

Kohl rabi
Radishes
French dressing

Caesar salad dressing

2 parts oil to 1 part lemon juice
Parmesan cheese
Chopped spring onion
Crushed clove garlic
Dry mustard
Sea salt
Black pepper
Worcester sauce

Hungarian cucumber salad

Cucumber (thin slices cut into rounds or longways) plus onion rings if liked
Dress with either

metric	*imperial*
3dl. boiling water	**½pt. boiling water**
1 tbsp. brown sugar	**1 tbsp. brown sugar**
3dl. wine vinegar	**½pt. wine vinegar**

or:
Soured cream seasoned with caraway seeds or anise or dill or curry powder to taste

13 Root vegetables recipes

Very earthy, for this chapter. Down among the spuds and turnips and other roots. High-class people used to turn up their noses at these splendid vegetables, probably because they're cheap and plentiful and loaded with nutriment. Well, I'm not so high-class, so I appreciate them! They contain lots of carbohydrate of course (the roots store sugar for the plant's new growth) but they're also rich in vitamins and minerals. Potatoes in particular are rich in vitamin C, and most people in this country get a good deal of their daily requirements from this source. Potatoes are so versatile I'll leave them till last, and start with carrots instead.

Carrots

These are among the aristocrats of vegetables, of course. You can't make a soup or stew without them, and you can use them for puddings as well as for savoury dishes (they improve a Christmas pudding remarkably), they're great in salads and they make super snacks for sweet-toothed kids. In fact, I think they're almost as indispensable as onions, so here's a recipe for using them together.

Use either tiny whole carrots, scrubbed, not peeled, or big carrots cut into strips, and tiny onions. Melt a generous dollop of butter in a heavy pan, and then add sugar (about a heaped teaspoonful), salt, pepper and the vegetables. Toss them around in the butter till they're well glazed. Now, add a minimum of water, just enough to prevent sticking, and put the lid on very tightly. Let the vegetables simmer slowly until they're tender. If the pan threatens to burn, a hint more boiling water is needed. Sprinkle with chopped parsley to serve. This dish looks as handsome as it tastes.

Kohl rabi

Now, let's look at a more exotic root, kohl rabi, which is really a sort of turnip. Just like ordinary turnip, it's delicious cooked to a purée, well drained and buttered, salted and peppered and served up with roast beef. Or you can roast it in pieces, under the beef, just like potatoes or parsnips. Or try it this way.

Young kohl rabi are parboiled in their skins, whole, then the skin is slipped off (and don't worry if some is left on—it doesn't matter a bit!) and the flesh is sliced thinly. Then arrange the vegetable in layers in a buttered dish, with chopped ham in between the layers. Alternate the layers with a rich cheese sauce. I make this as a classic white sauce, adding lots of grated cheese and a dollop of mustard. Finish with a sauce layer, and sprinkle the top with brown breadcrumbs mixed with grated cheese, garnish with a couple of tomato slices and sprinkle with parsley for pretty, and bake in a moderately hot oven—Regulo 6/7, 205°C/215°C, or 400°F—for about half an hour. Very suppery this. It works just as well with celeriac or turnips or potatoes, by the way. I always add onions when I make it with potatoes, though.

Salsify and scorzonera

And here is another exotic root, salsify. It is very similar to scorzonera—snake plant, the ancients called that, because they reckoned it cured snake-bite. There's no evidence it ever did, mind you! I suspect that was a bit of sympathetic magic because it looks so snaky, being long, black and tapery. Since we have no snakes, we'll eat it like salsify, which is sometimes called vegetable oyster, by the way, because it has a faintly fishy taste. It's an acquired taste, I admit—which I have. I like both salsify and scorzonera best scrubbed lightly—and all the flavour is in the skin, so please *be* light—baked in the oven in foil with butter, after which the skins can be slipped off to serve. The sooner you cook them after they're dug up the better and the sooner you eat them after they're cooked the better. They lose heart very quickly! You can use a white sauce if you like, but I think they're best simply with melted butter.

I'm told you can add the buds of scorzonera or salsify flowers to an omelette. You wash them, dry them, fry them till they're brown, and then add them to the egg mixture. Sounds pretty, because the flowers open as they cook. I couldn't try it, though—no flowers available! However, I have fried other vegetables flowers very successfully, notably the bright yellow marrow flower. Dip it in well-beaten egg, dredge lightly with seasoned flour, and deep fry very quickly till golden brown. Eat at once, and they're lovely with escalopes of veal. And at the price veal is, it deserves a bit of fancy dressing when you serve it!

Turnip tops

Let's return to roots, and to economy. Sometimes you can use the bits that usually get chucked away. For example, turnip tops. Try them Chinese style. Pour boiling water over them to make them softer, and then shred them as fine as you can. Then melt oil-and-butter and soy sauce to taste in a heavy pan and stir-fry the greens in it for a couple of minutes. Just enough to heat them through, really. Then add some fried nuts—cashews are great, but you can use almonds or even peanuts.

Turnip tops, like dark green cabbage leaves, make good soup too. Shred very very finely, and chop an onion very finely. Cook the onion in oil and butter till transparent, then add a half and half mixture of chicken stock and milk. Season lightly with marjoram, and last of all, stir in as much shredded greens as necessary to make the soup fairly bulky, but not solid! Eat at once, before the greens can lose their crispness, with bread croûtons tossed on top of each bowl. Filling, nutritious and very very economical!

Silver seakale beet

Speaking of leafy tops of vegetables, what about silver seakale beet? You can cook the leaves like spinach, but it's the stalks that I like best. Strip away the mid-ribs from each leaf, and wash them well. Then tie them into loose bundles and poach them. Don't overcook or the ribs will go tough, and lose their flavour. Eat them like asparagus—with melted butter or a white sauce or a cheese sauce, or even a vinaigrette sauce.

Potatoes

And now to our dear old friends the spuds. I can get quite lyrical about them! Boil them with mint or fry them raw in butter and nutmeg, or parboil and then fry them with onions, or mash them with cream, or bake them in their jackets—well, you all know that! So here's a different way with them. Potato Latkes, and I can tell you when I made these on the programme, the cameramen and the electricians and the riggers and even the producer were queueing up to get their share—they smell marvellous while they are cooking and taste even better than they smell!

Use grated raw potato, and add grated raw onion—I weep bitter tears when I make this dish, but it's worth it! The potato must be left standing in a colander for about an hour, to let the excess liquid drain out. It will blacken a little of course, but that doesn't matter. Then press the rest of the liquid out, and mix the onion and potato together, with beaten eggs, lots of salt and pepper, and if it's a bit sloppy a little flour. As ever, I can't give too exact quantities because the size of vegetables vary—but for about 1kg. (2lbs.) of potatoes and 1 large onion you'll need 2 or 3 eggs. Now, in a heavy pan heat oil till it smokes and then drop in spoonfuls of the mixture. They'll fluff up, but also look ragged. Never mind—that adds to their charm! Turn them and flatten them a little if necessary, and when they are golden brown, drain on kitchen paper and eat hot. They couldn't be more fattening, and couldn't be more delicious! For a variation, the mixture can be piled into a buttered oven dish and baked in a hot oven (Regulo 7, 215°C, 425°F) for about 45 minutes, or till it's firm and golden brown. This way is rather heavier but just as delicious, and makes a change from roast potatoes with the Sunday joint.

Carrots

Carrots
Tiny onions
Butter
Sugar
Salt and pepper
Chopped parsley

Cook altogether over a low light for about half an hour.

Kohl rabi

SERVES 6

metric	*imperial*
2 or 3 large kohl rabi	**2 or 3 large kohl rabi**
125g. chopped ham	**¼lb. chopped ham**
9dl. cheese sauce	**1½pt. cheese sauce**
Breadcrumbs	**Breadcrumbs**
Tomato slices	**Tomato slices**
Chopped parsley	**Chopped parsley**

Bake in moderately hot oven (Regulo 6/7; 205°C/215°C; 400°F/415°F) for about ½ hour.

Turnip tops

Turnip tops
Oil and butter
Nuts, fried
Soy sauce

Soak in hot water, shred. Cook in oil and butter with soy sauce, till just crisp. Add nuts. Eat at once. (Cooks in 5-10 minutes.)

Potato latkes

MAKES 1-1½ DOZEN CAKES

metric	*imperial*
About 1Kg. potatoes, grated	**About 2lbs. potatoes, grated**
1 large onion, grated	**1 large onion, grated**
Salt and pepper	**Salt and pepper**
2 beaten eggs	**2 beaten eggs**
Flour	**Flour**

Drain potatoes and onions. Mix and season well. Add beaten egg to bind and a little flour if the mixture is too wet, and fry in spoonfuls in very hot oil. Drain on kitchen paper and eat at once.

14 A vegetable banquet

For the final programme in the series, I devised a mini-banquet. And because it was for a gardener (and I know these gardeners. Tell 'em dinner's at seven and you're lucky if you see 'em before half past eight!) everything I made was the sort of dish that could sit and wait without damage. So, the first course was a cold soup.

I grated 2 large cucumbers and 1 large onion roughly, and then sweated it all in melted butter in a heavy pan. It takes a little while to reduce the vegetables to a mush—it's more like melting them really. Then I added a mixture of milk and strong chicken broth. I make my broth the old-fashioned way, with your actual chicken, but a cube will do! Cook the mixture in an open pan slowly (it'll boil over if you rush it!) until the onion is cooked through—a matter of 10 to 15 minutes I find is usually enough. Then put it in the blender for a second or two to get it nice and smooth, and chill it thoroughly in the fridge. Serve it with a dollop of soured cream on top and chopped watercress to garnish.

The second course is really very simple but looks very fancy. In a pre-baked pie shell made of *pâte brisée* (see page 53) I arranged lots of different cooked vegetables in neat patterns. Fried onions, glazed baby carrots, cooked beans, peas, chopped celery—whatever there was available. I made two; for one I used a very well flavoured aspic to cover the vegetables, made of gelatine, strong chicken stock and sherry. When it set it looked beautiful. A super show-offy party dish! For the other I used an unbaked pie shell, and poured over the top of the vegetables arranged as before, a savoury custard mixture—three eggs beaten into 3dl. (½pt.) milk with a little added cream to be luxurious. I seasoned the top (a little nutmeg, salt and pepper) and baked it in a moderate oven (Regulo 4/5, 175°C/190°C, 350°F/375°F) till it was brown and set. This dish can be eaten hot or cold. If you want to be fancy, call it a *quiche des legumes*. Or be plain, like me, and call it a vegetable tart.

Then I made one of my favourites—stuffed cabbage! Cabbage leaves are carefully removed from the parent plant and put in very hot water for a while to soften them. The stuffing I use is minced beef mixed together with grated onion, seasoned with salt and pepper, garlic and the herb of your choice. I used marjoram. I also added a handful of uncooked rice and bound the lot together with an egg.

Each leaf is then given a dollop of meat and rice and packed into a neat little parcel, with the corners folded in firmly. Arrange the parcels tidily in an oven dish so that they really are packed tight, and then pour over the top some very concentrated chicken or beef stock. Some people add tomato purée, brown sugar and a tablespoonful of vinegar at this stage, to make the dish sweet and sour. I prefer it this plainer way. Bake the parcels, covered, in a warm to moderate oven (Regulo 3/4, 165°C/

175°C, 325°F/350°F) for about 1 hour or even 2, to make sure the rice is cooked through. They're marvellously filling, taste as good as they look, and if you've any left over, they can be frozen very successfully.

Cucumber soup

SERVES 6

metric	*imperial*
2 cucumbers	**2 cucumbers**
1 large onion	**1 large onion**
3dl. milk	**½pt. milk**
3dl. chicken broth	**½pt. chicken broth**
Soured cream	**Soured cream**
Watercress	**Watercress**

Grate cucumber and onion, cook in butter for 10-15 minutes till soft. Add milk and broth, cook for another 10-15 minutes. Blend, chill, serve with soured cream and chopped watercress.

Aspic tart

SERVES 8

metric	*imperial*
***Baked* pie shell (pâte brisée— see page 53)**	***Baked* pie shell (pâte brisée— see page 53)**
Onions	**Onions**
Baby carrots	**Baby carrots**
Beans	**Beans**
Peas	**Peas**
Celery	**Celery**

Cook lightly and arrange in triangles.

Glaze with aspic made of:

12-15g. gelatine	**½oz. gelatine**
6dl. chicken stock	**1pt. chicken stock**
Sherry to taste	**Sherry to taste**

Serve cold.

Vegetable quiche

SERVES 8

metric	*imperial*
***Un*baked pie shell (pâte brisée— see page 53)**	***Un*baked pie shell (pâte brisée— see page 53)**
Vegetables as for aspic tart	**Vegetables for aspic tart**
3 eggs	**3 eggs**
3dl. milk	**½pt. milk**
1.5dl. cream	**¼pt. cream**
Salt and pepper to taste	**Salt and pepper to taste**

Beat well, pour over vegetables and bake at Regulo 4/5, 175°C/190°C, 350°F/375°F until firm and golden brown (approximately 15-20 minutes).

Stuffed cabbage

SERVES 6

metric	*imperial*
6 large cabbage leaves	**6 large cabbage leaves**
450g. minced beef	**1lb. minced beef**
2 onions	**2 onions**
80g. uncooked rice	**3ozs. uncooked rice**
Salt and pepper	**Salt and pepper**
Garlic and herbs	**Garlic and herbs**
1 egg	**1 egg**
Chicken stock	**Chicken stock**

Soak leaves in hot water till pliable. Mix beef, onions, seasoning and uncooked rice, bind with egg and put a small dessertspoonful on each leaf. Parcel firmly, pack tightly in oven dish. Add chicken stock, and bake covered at Regulo 3/4, 165°C/175°C, 325°F/350°F for 1½-2 hours, until tender.

You can alternate the stuffings for the cabbage leaves. Try the following alternatives:

Breadcrumbs, minced ham and sage
Minced lamb, pearl barley, bouquet garni
Minced lamb, lentils and split peas, bay leaf
Minced liver, oatmeal, *herbes de Provence*.

So there you are—the results of seven weeks of vegetable cookery for a television series. Here's hoping you enjoyed watching the programmes as much as I enjoyed cooking for them—and that you find this record of the recipes useful!